Instructor's Manual
and Printed Test Bank

College Algebra

SECOND EDITION

Instructor's Manual
and Printed Test Bank

College Algebra
SECOND EDITION

BITTINGER
BEECHER

Donna DeSpain

Addison-Wesley Publishing Company
Reading, Massachusetts • Menlo Park, California • New York
Don Mills, Ontario • Wokingham, England • Amsterdam • Bonn
Sydney • Singapore • Tokyo • Madrid • San Juan • Milan • Paris

1 2 3 4 5 6 7 8 9 10-BK-96959493

TABLE OF CONTENTS

I. ALTERNATE TESTS, FORMS A, B, C, D, E, and F 1

There are six alternate test forms for each chapter and the final
examination. Alternate Test Forms A, B, C, and D are equivalent in
length and difficulty. Synthesis questions occur at the end of
Test Forms A, B, C, and D and are separated from the rest of the test
by a solid line. Synthesis questions are meant to be more challenging
like those problems found in the last part of each exercise set. The
synthesis questions have been placed at the end to make it easy to omit
them if the instructor wishes to do so.

Alternate Test Form E adds variety in style of questions and in the
objectives tested. In most chapters Test Form E is more difficult
than Test Forms A, B, C, D, and F.

With the exception of mathematical induction, all questions on Test
Form F are multiple choice. Effort was made to make the wrong answers
as logically wrong as possible. In most cases answers were
constructed to avoid students doing backward reasoning.

Special thanks are extended to Randy Forchee and Judy Salmon of Daniel
Webster College for their tremendous job of checking the manuscript.
Thanks also to Peggy Carter for her excellent typing and organizational
skills.

Instructor's Manual
and Printed Test Bank

College Algebra
SECOND EDITION

NAME _____

CLASS _____ SCORE _____ GRADE _____

	ANSWERS

Consider the numbers:

$$12, \ -\frac{4}{3}, \ -\sqrt[4]{5}, \ 0, \ 0.58, \ 8\frac{1}{2}, \ \sqrt{50}, \ -5.4, \ 7.0\overline{3}, \ -11.$$

1. Which are whole numbers?

2. Which are irrational numbers?

3. Which are real numbers?

4. Which are rational numbers?

5. Which are natural numbers?

6. Which are integers?

Compute.

7. $-11 + |-11|$

8. $-3.9 + 7.6$

9. $(-15)(-3)$

10. $5 - (-14)$

11. $\dfrac{21}{-7}$

12. $\dfrac{32 \div 2^3 - 5^2 \cdot 3^2}{4(5 - 13) + 6 \cdot 5}$

Convert to decimal notation.

13. 5.401×10^{-3}

14. 5.8×10^4

15. $3.15 \ E \ 5$

16. $6.17 \ E \ -3$

ANSWERS

1. _____
2. _____
3. _____
4. _____
5. _____
6. _____
7. _____
8. _____
9. _____
10. _____
11. _____
12. _____
13. _____
14. _____
15. _____
16. _____

NAME _____

ANSWERS	

Convert to scientific notation.

17. 0.0000217

18. 593.21

17. _____

18. _____

Compute. Write scientific notation for each answer.

19. $(6.8 \times 10^4)(5.1 \times 10^{-15})$

20. $\dfrac{4.76 \times 10^{-18}}{5.6 \times 10^{-12}}$

19. _____

20. _____

Simplify.

21. $(5x^3y^{-3})(-2x^4y^{-5})$

22. $\dfrac{14ab^4c^{-2}}{21a^{-3}b^{-4}c^5}$

21. _____

22. _____

23. $\sqrt[3]{-125}$

24. $\sqrt[4]{81}$

23. _____

25. $\left(\sqrt{7} - \sqrt{3}\right)\left(\sqrt{7} + \sqrt{3}\right)$

26. $\dfrac{\dfrac{1}{x^2} - \dfrac{1}{y^2}}{\dfrac{x^2 + 2xy + y^2}{xy}}$

24. _____

25. _____

27. $(4a^3 + 3b)^2$

26. _____

28. $(5x^3y + 4x^2 - 2y^2 + 3) - (-2x^3y + 3xy - y^2 - 5)$

27. _____

28. _____

29. $(2x - 1)^3$

29. _____

30. Write an expression containing a single radical:

$$\dfrac{\sqrt[3]{(a + b)^2} \cdot \sqrt{a + b}}{\sqrt[4]{(a + b)^3}}.$$

30. _____

NAME _____

31. Convert to radical notation: $c^{-3/4}$.

Factor.

32. $45x^2 + 60x + 20$

33. $a^3 + 216$

34. $b^5 - 4b^3c^2$

35. $12m^4 - 26m^2 - 30$

36. $729 - 64a^6$

37. $5x + 10y - x^2 - 2xy$

38. $a^9 - 27a^3b^6$

39. Divide and simplify: $\dfrac{x^2 - x - 6}{x^2 - 6x + 9} \div \dfrac{(x + 2)^3}{x^2 - 9}$.

40. Add and simplify: $\dfrac{y}{y^2 + 4y - 21} + \dfrac{5}{y^2 - 9}$.

41. Rationalize the denominator: $\dfrac{8 + \sqrt{x}}{8 - \sqrt{x}}$.

42. Multiply: $(a^x + 3a^{-x})^3$.

43. Rationalize the numerator: $\dfrac{8 + \sqrt{x}}{8 - \sqrt{x}}$.

ANSWERS
31. _____
32. _____
33. _____
34. _____
35. _____
36. _____
37. _____
38. _____
39. _____
40. _____
41. _____
42. _____
43. _____

NAME _____

ANSWERS	
44. _____	44. Factor: $a^{1/2}b^{2/3} - a^{-1/2}b^{5/3}$.
45. _____	45. The diagonal of a square has length $10\sqrt{2}$. Find the length of a side of the square.
46. _____	46. Factor and simplify: $\dfrac{2x^2(3x - 1)^{1/2} - 3x^2(3x - 1)^{-1/2}}{[(3x - 1)^{1/3}]^3}$.
47. _____	47. Change $50 \dfrac{ft}{sec}$ to $\dfrac{mi}{hr}$.
48. _____	48. Multiply: $(x^a - y^b)^3$.
49. _____	49. Simplify: $\sqrt{2 + x} + \dfrac{1}{\sqrt{2 + x}}$.

NAME _____

CLASS _____ SCORE _____ GRADE _____

Consider the numbers:

$5.\overline{7}$, $-\sqrt[3]{9}$, $\frac{6}{7}$, -0.154, 13, $\sqrt{18}$, 0, $-4\frac{2}{3}$, -5, 4.1.

1. Which are whole numbers?

2. Which are irrational numbers?

3. Which are real numbers?

4. Which are rational numbers?

5. Which are natural numbers?

6. Which are integers?

Compute.

7. $-\frac{6}{5}\left(-\frac{2}{3}\right)$

8. $19 - 35$

9. $5.2 + (-7.5)$

10. $-3 + |-15|$

11. $-\frac{10}{3} \div \frac{4}{5}$

12. $\dfrac{75 \div 5^2 - 3^2 \cdot 4^2}{3(6 - 19) + 2 \cdot 7}$

Convert to decimal notation.

13. 6.13×10^5

14. 4.312×10^{-2}

15. 7.92 E 4

16. 9.02 E -5

ANSWERS
1. _____
2. _____
3. _____
4. _____
5. _____
6. _____
7. _____
8. _____
9. _____
10. _____
11. _____
12. _____
13. _____
14. _____
15. _____
16. _____

NAME _____

ANSWERS	
17. _____	Convert to scientific notation.
	17. 0.0000032 18. 4374.52
18. _____	
	Compute. Write scientific notation for each answer.
19. _____	19. $(6.2 \times 10^{-3})(8.1 \times 10^{-6})$ 20. $\dfrac{1.26 \times 10^3}{2.8 \times 10^{-8}}$
20. _____	
	Simplify.
21. _____	21. $(3x)^4(2x)^3$ 22. $\dfrac{16x^2y^3z^{-4}}{24x^{-1}y^4z^{-3}}$
22. _____	
	23. $\sqrt[3]{-1000}$ 24. $\sqrt[5]{243}$
23. _____	
24. _____	25. $\left(\sqrt{6} + \sqrt{3}\right)\left(\sqrt{6} - \sqrt{3}\right)$ 26. $\dfrac{\dfrac{x^2 - y^2}{2xy}}{\dfrac{x + y}{y}}$
25. _____	
26. _____	27. $(5x^2 - 4y^3)^2$
27. _____	28. $(5x^2y + 2xy^2 - 5x - 2y) + (4x^3 - 3xy^2 - 9x - 6)$
28. _____	29. $(5y - 4)^3$
29. _____	
30. _____	30. Write an expression containing a single radical: $\dfrac{\sqrt[5]{c + d} \cdot \sqrt[4]{(c + d)^3}}{\sqrt[3]{(c + d)^2}}.$

	ANSWERS

31. Convert to radical notation: $x^{4/5}$.

31. _____

Factor.

32. _____

32. $24x^2 - 120x + 150$ 33. $b^3 - 512$

33. _____

34. $28x^7 - 7x^3z^2$ 35. $24a^4 + 30a^2 - 126$

34. _____

36. $64 + 27a^6$ 37. $8x - 4y - 2x^2 + xy$

35. _____

38. $a^9 - 9a^3b^4$

36. _____

39. Multiply and simplify: $\dfrac{3x^2 - 13x - 10}{2x^2 + 14x} \cdot \dfrac{x^2 - 49}{3x^2 - x - 2}$.

37. _____

40. Subtract and simplify: $\dfrac{3x}{3x - 5y} - \dfrac{5y}{5y - 3x}$.

38. _____

39. _____

41. Rationalize the denominator: $\dfrac{4 - \sqrt{x}}{4 + \sqrt{x}}$.

40. _____

42. Multiply: $(3a^x - 4b^{-x})^3$.

41. _____

43. Rationalize the numerator: $\dfrac{4 - \sqrt{x}}{4 + \sqrt{x}}$.

42. _____

43. _____

NAME _____

ANSWERS	
44. _____	44. Factor: $a^{7/4}b^{-2/3} + a^{3/4}b^{1/3}$.
45. _____	45. An airplane is flying at an altitude of 4200 ft. The slanted distance directly to the airport is 10,300 ft. How far horizontally is the airplane from the airport?
46. _____	46. Factor and simplify: $\dfrac{x(x-3)^{1/2} - (x+3)(x-3)^{-1/2}}{x-3}$
47. _____	47. Change $80 \dfrac{lb}{ft^3}$ to $\dfrac{ton}{yd^3}$.
48. _____	48. Multiply: $(2a + b + 3c)^3$.
49. _____	49. Simplify: $\dfrac{(a-b)^3 + b^3}{b}$.

NAME _____

CLASS _____ *SCORE* _____ *GRADE* _____

ANSWERS

Consider the numbers:

$$-\sqrt{39},\ 415,\ 8\tfrac{1}{2},\ 0.\overline{21},\ -\sqrt[3]{4},\ -19,\ 0,\ \sqrt{13},\ -\tfrac{5}{8},\ -9.1.$$

1. Which are whole numbers?

2. Which are irrational numbers?

3. Which are real numbers?

4. Which are rational numbers?

5. Which are natural numbers?

6. Which are integers?

Compute.

7. $(-7.4) \times 8$

8. $\dfrac{-35}{-7}$

9. $15 + |-7|$

10. $-11.07 - 15.32$

11. $\dfrac{5}{8} + \left(-\dfrac{3}{5}\right)$

12. $\dfrac{5(19 - 3^2) - 4^2 \cdot 6}{7(5 - 13) + 3 \cdot 8}$

Convert to decimal notation.

13. 9.37×10^5

14. 6.12×10^{-4}

15. $4.09\ \mathrm{E}\ 7$

16. $8.53\ \mathrm{E}\ -2$

1. _____

2. _____

3. _____

4. _____

5. _____

6. _____

7. _____

8. _____

9. _____

10. _____

11. _____

12. _____

13. _____

14. _____

15. _____

16. _____

NAME _____

ANSWERS
17. _____
18. _____
19. _____
20. _____
21. _____
22. _____
23. _____
24. _____
25. _____
26. _____
27. _____
28. _____
29. _____
30. _____

Convert to scientific notation.

17. 0.000537

18. 829.15

Compute. Write scientific notation for each answer.

19. $(5.9 \times 10^3)(9.2 \times 10^{-11})$

20. $\dfrac{1.075 \times 10^{-15}}{4.3 \times 10^{-7}}$

Simplify.

21. $(-5x^3)^4$

22. $\dfrac{8a^2b^3c^{-4}}{32ab^{-5}c}$

23. $\sqrt[3]{-216}$

24. $\sqrt[4]{256}$

25. $\left(\sqrt{7} - 2\right)^2$

26. $\dfrac{a - \dfrac{a}{b}}{a^2 - \dfrac{a^2}{b}}$

27. $(2x - 3)(3x^2 - x + 4)$

28. $(9x^4 + 2x^3y - 4xy^2 - 11) - (3x^3 + 5xy^2 - 2x + 4)$

29. $(4a - 3)^3$

30. Write an expression containing a single radical:

$$\dfrac{\sqrt{a - b} \cdot \sqrt[5]{(a - b)^4}}{\sqrt[4]{(a - b)^3}}.$$

NAME _____

	ANSWERS

31. Convert to radical notation: $a^{-2/3}$.

31. _____

Factor.

32. $75x^2 + 30x + 3$ 33. $a^3 + 729$

32. _____

33. _____

34. $6x^2y^4 - 6x^2z^2$ 35. $18m^4 + 51m^2 - 9$

34. _____

36. $64 + 216b^6$ 37. $4x - 20y + x^2 - 5xy$

35. _____

38. $a^{11} + 8a^2b^9$

36. _____

39. Divide and simplify: $\dfrac{x^2 + 11x + 28}{x^2 + 2x - 3} \div \dfrac{x^2 + 8x + 16}{x^2 + 4x - 5}$.

37. _____

40. Add and simplify: $\dfrac{3x - 1}{5x^2 + 18x - 8} + \dfrac{6}{x^2 - x - 20}$.

38. _____

39. _____

41. Rationalize the denominator: $\dfrac{5 + \sqrt{x}}{5 - \sqrt{x}}$.

40. _____

42. Multiply: $(2a^x - 4a^{-x})^3$.

41. _____

43. Rationalize the numerator: $\dfrac{5 + \sqrt{x}}{5 - \sqrt{x}}$.

42. _____

43. _____

NAME _____

ANSWERS

44. _____

45. _____

46. _____

47. _____

48. _____

49. _____

44. Factor: $a^{8/5}b^{-1/2} - a^{-2/5}b^{1/2}$.

45. A baseball diamond is actually a square 90 ft on a side. A catcher fields a bunt along the third-base line 15 ft from home plate. How far would the catcher have to throw the ball to first base?

46. Factor and simplify: $\dfrac{x^2(x^2 - 5)^{-1/2}x - (3x)(x^2 - 5)^{1/2}}{[(x^2 - 5)^{1/2}]^2}$.

47. Change 2400 $\dfrac{g}{L}$ to $\dfrac{cg}{mL}$.

48. Multiply: $(a^{2x+y} \cdot a^{x-y})^4$.

49. Factor: $x^{16} - 625$.

NAME _____

CLASS _____ *SCORE* _____ *GRADE* _____

ANSWERS

Consider the numbers:

$$-5\frac{3}{4}, \ 15.6, \ \sqrt{80}, \ -3.14, \ -19, \ 5.\overline{27}, \ -\sqrt[4]{52}, \ 0, \ \frac{11}{3}.$$

1. Which are whole numbers?

2. Which are irrational numbers?

3. Which are real numbers?

4. Which are rational numbers?

5. Which are natural numbers?

6. Which are integers?

Compute.

7. $-5.6 + (-9.3)$

8. $-25.5 \div -0.3$

9. $\frac{17}{9} - \left(-\frac{2}{3}\right)$

10. $-815 + 409$

11. $|-17| - |-12|$

12. $\dfrac{14 \cdot 3^2 - 52 \div 2^2}{4(5 + 3^2) - 7 \cdot 3}$

Convert to decimal notation.

13. 2.175×10^{-2}

14. 9.1×10^3

15. $5.09 \ E \ 6$

16. $1.32 \ E \ -4$

1. _____

2. _____

3. _____

4. _____

5. _____

6. _____

7. _____

8. _____

9. _____

10. _____

11. _____

12. _____

13. _____

14. _____

15. _____

16. _____

NAME _____

ANSWERS	Convert to scientific notation.

17. _____

17. 0.00132   18. 574.9

18. _____

Compute. Write scientific notation for each answer.

19. _____

19. $(7.3 \times 10^{-6})(5.3 \times 10^{14})$ 20. $\dfrac{1.76 \times 10^{-4}}{3.2 \times 10^{-8}}$

20. _____

Simplify.

21. _____

21. $(-6x^3y^{-2})(2x^{-4}y)$ 22. $(-3c^{-2}d^3)^{-2}$

22. _____

23. $\sqrt[3]{-64}$ 24. $\sqrt[5]{32}$

23. _____

25. $\left(1 + \sqrt{5}\right)^2$ 26. $\dfrac{\dfrac{1}{x^2} - \dfrac{1}{y^2}}{\dfrac{2x^2 + 3xy + y^2}{xy}}$

24. _____

25. _____

27. $(x^2 + ab)(x^2 - ab)$

26. _____

27. _____

28. $(-4pq^2 + 7pq - 11p + 12q) + (11pq^2 - 14q^2 + 7q - 9)$

28. _____

29. $(3x - 2)^3$

29. _____

30. Write an expression containing a single radical:

30. _____

$\dfrac{\sqrt[6]{(x + y)^5} \cdot \sqrt[4]{x + y}}{\sqrt{x + y}}$.

NAME _____

	ANSWERS
31. Convert to radical notation: $y^{3/8}$.	31. _____
Factor.	32. _____
32. $18x^2 - 48x + 32$ 33. $c^3 - 1000$	33. _____
34. $a^5 - 16a^3b^2$ 35. $16p^4 + 26p^2 - 12$	34. _____
36. $512 - 27a^9$ 37. $6x + 24y - x^2 - 4xy$	35. _____
38. $8a^6b^2 - b^8$	36. _____
39. Multiply and simplify: $\dfrac{x^2 - x - 30}{x^2 - 1} \cdot \dfrac{x^2 - x - 2}{(x + 5)^3}$.	37. _____
40. Subtract and simplify: $\dfrac{5y}{y^2 - 7y + 6} - \dfrac{3y}{y^2 - 4y - 12}$.	38. _____
	39. _____
41. Rationalize the denominator: $\dfrac{6 - \sqrt{x}}{6 + \sqrt{x}}$.	40. _____
42. Multiply: $(5x^t + x^{-t})^3$.	41. _____
43. Rationalize the numerator: $\dfrac{6 - \sqrt{x}}{6 + \sqrt{x}}$.	42. _____
	43. _____

NAME _____

ANSWERS	
44. _____	44. Factor: $a^{-1/3}b^{21/8} - a^{2/3}b^{5/8}$.
45. _____	45. How long must a wire be to reach from the top of a 16-m telephone pole to a point on the ground 9 m from the foot of the pole?
46. _____	46. Factor and simplify: $$\frac{5x^2(2x + 1)^{-1/2} - x^3\left(\frac{2}{3}\right)(2x + 1)^{1/2}(3)}{[(2x + 1)^{1/4}]^4}.$$
47. _____	47. Change $40\ \frac{kg}{m}$ to $\frac{g}{cm}$.
48. _____	48. Convert to fractional notation: $4.\overline{24}$.
49. _____	49. Factor: $y^2 - \frac{15}{64} - \frac{1}{4}y$.

NAME _____

CLASS _____ SCORE _____ GRADE _____

	ANSWERS

1. In the following list of numbers, circle each
    irrational number.

    16.29, $4\frac{1}{3}$, 0, $\sqrt{50}$, 19.1, $\frac{13}{5}$, -47, $-\frac{2}{3}$, π, $\sqrt[3]{15}$, $\sqrt{169}$,

    0.043333 . . . (Numeral repeats),

    17.12112111211112 . . . (Numeral does not repeat)

2. Find $-(x^3 + 2x^2 - 9)$ when $x = -1$.

3. What property is illustrated by this sentence?

    $-17(ab) = (-17a)b$

Compute.

4. $|-7.9| + |14.2|$ 5. $7 + (-18)$ 6. $5(-7)(-4)$

7. $-19 - (-3)$ 8. $\dfrac{-85}{-5}$ 9. $\dfrac{2}{3} \div (-3)$

10. Convert to decimal notation: 9.7×10^{-3}.

11. Convert to scientific notation: 5,073,000.

Simplify.

12. $(-8x^2y^{-3})(-2xy^{-2})$ 13. $\left(5\sqrt{a} - \sqrt{2}\right)\left(5\sqrt{a} + \sqrt{2}\right)$

14. $\sqrt[3]{343}$ 15. $\dfrac{16r^{-9}s^8t^4}{20r^{-5}st^{-5}}$ 16. $\dfrac{\dfrac{a}{a-b} - \dfrac{b}{a+b}}{\dfrac{ab}{a^2 - b^2}}$

17. $(2x^3 - y^2)^2$ 18. $(2a - 7b)^3$ 19. $\sqrt[5]{-1}$

20. $(2x^5 - 3x^2y - 5y^2 + 2) - (-x^5 - 5xy + y^2 - 9)$

ANSWERS

1. See answer at left. _____

2. _____

3. _____

4. _____

5. _____

6. _____

7. _____

8. _____

9. _____

10. _____

11. _____

12. _____

13. _____

14. _____

15. _____

16. _____

17. _____

18. _____

19. _____

20. _____

NAME _____

21. _____

22. _____

23. _____

24. _____

25. _____

26. _____

27. _____

28. _____

29. _____

30. _____

31. _____

32. _____

33. _____

34. _____

35. _____

21. Compute. Write scientific notation for the answer.

$$\frac{(2.9 \times 10^{-7})(1.29 \times 10^{4})}{4.3 \times 10^{8}}$$

Factor.

22. $48x^{4} - 3$ 23. $y^{3} + 0.027$ 24. $3x^{2} - 25x - 18$

25. $8(3a - b)^{2} + 2(3a - b) - 21$

26. Write an expression containing a single radical:

$$\frac{\sqrt[5]{(b + 3)^{2}} \cdot \sqrt[4]{b + 3}}{\sqrt[10]{(b + 3)^{3}}}.$$

27. Convert to radical notation: $z^{-9/11}$.

28. Convert to exponential notation and simplify:

$$\sqrt[8]{\frac{x^{24}y^{32}}{z^{40}}}.$$

29. Divide and simplify: $\dfrac{x^{2} - 6x - 27}{x^{2} + 5x - 14} \div \dfrac{x^{2} + 3x}{x^{2} - 49}.$

30. Subtract and simplify: $\dfrac{8}{a + b} - \dfrac{a - b}{a^{2} - b^{2}}.$

31. Rationalize the denominator: $\dfrac{4\sqrt{a} - 5\sqrt{b}}{\sqrt{a} + \sqrt{b}}.$

32. Factor and simplify: $3x^{-1/4}y^{2/3} + 5x^{3/4}y^{-1/3}.$

33. A 6-m ladder is leaning against a building. The bottom of the ladder is 4 m from the building. How high is the top of the ladder?

34. Change $\dfrac{\$64}{\text{day}}$ to $\dfrac{\cancel{c}}{\text{hr}}$.

35. Multiply: $(y^{b} + z^{c})(y^{b} - z^{c}).$

NAME _____

CLASS _____ SCORE _____ GRADE _____

Write the letter of your response on the answer blank.

ANSWERS

1. Add: $|-25| + (-9)$.

   a) -34 b) 16 c) 34 d) -16

   1. _____

2. Subtract: $-7 - (-13)$.

   a) -20 b) -6 c) 20 d) 6

   2. _____

3. Multiply: $6(-5)(-7)$.

   a) -210 b) -20 c) 210 d) 180

   3. _____

4. Divide: $\dfrac{-3.6}{-4}$.

   a) 9 b) -0.9 c) 0.9 d) -7.6

   4. _____

5. Find $-(3x^2 - 4x + 8)$ when $x = -2$.

   a) -28 b) 6 c) -4 d) 28

   5. _____

6. Convert to decimal notation: 8.217×10^{-4}.

   a) $82,170$ b) 0.0008217 c) 8217 d) 8.217

   6. _____

7. Convert to scientific notation: 0.0000263.

   a) 0.263×10^{-3} b) 2.63×10^{-5}

   c) 2.63×10^{-4} d) 2.63×10^{4}

   7. _____

8. Simplify: $(3x^2y^{-3})(-5x^{-7}y)$.

   a) $\dfrac{15}{x^5y^2}$ b) $\dfrac{-15}{x^5y^4}$ c) $\dfrac{-2x^2}{y^4}$ d) $\dfrac{-15}{x^5y^2}$

   8. _____

9. Simplify: $\sqrt[5]{-243}$.

   a) -3 b) -9 c) 3 d) 7

   9. _____

10. Simplify: $\dfrac{24x^3y^4z^{-1}}{36xy^{-2}z^4}$.

    a) $\dfrac{2x^2y^6}{3z^5}$ b) $\dfrac{2xy^2}{3z^3}$ c) $\dfrac{2x^4y^6}{3z^2}$ d) $\dfrac{2x^2y^2}{3z^5}$

    10. _____

19

ANSWERS	11. Multiply and simplify: $(4y^3 - x^2)^2$.
11. _____	a) $8y^6 - 8x^2y^3 + x^4$ b) $16y^6 - x^4$
	c) $16y^9 + x^4$ d) $16y^6 - 8x^2y^3 + x^4$

11. Multiply and simplify: $(4y^3 - x^2)^2$.

a) $8y^6 - 8x^2y^3 + x^4$　　　　b) $16y^6 - x^4$

c) $16y^9 + x^4$　　　　d) $16y^6 - 8x^2y^3 + x^4$

12. Multiply and simplify: $(3x - 5)^3$.

a) $27x^3 - 125$　　　　b) $27x^3 - 135x^2 + 225x - 125$

c) $27x^2 - 30x + 125$　　　　d) $27x^3 - 15x^2 + 15x - 125$

13. Multiply and simplify: $\left(\sqrt{3}x + y + \sqrt{5}\right)\left(\sqrt{3}x + y - \sqrt{5}\right)$.

a) $3x^2 + 6\sqrt{3}y - 5$　　　　b) $9x^2 + 6xy + y^2 - 5$

c) $3x^2 + y^2 - 5$　　　　d) $3x^2 + 2\sqrt{3}xy + y^2 - 5$

14. Simplify: $\dfrac{1 - \dfrac{3}{2x}}{x - \dfrac{9}{4x}}$.

a) $\dfrac{2}{2x + 3}$　　　　b) $\dfrac{(3 - 2x)(4x - 1)}{9}$

c) $\dfrac{2(2x - 3)}{4x - 9}$　　　　d) $\dfrac{4}{2x - 3}$

15. Subtract and simplify:

$(7x^2 - 5xy + 2y^2 - 4) - (3x^2 + 5xy - y^2 + 3)$.

a) $4x^2 - 10xy + 3y^2 - 7$　　　b) $4x^2 - y^2 - 1$

c) $4x^2 - 10xy + y^2 - 1$　　　d) $4x^2 - 10xy - y^2 - 7$

16. Factor: $5y^3 - 40$.

a) $5(y + 2)(y^2 - 2y - 4)$　　b) $5(y + 2)(y^2 - 10y + 8)$

c) $5(y - 2)(y^2 + 2y + 4)$　　d) $5(y - 2)(y^2 - 10y - 4)$

17. Factor: $5r^2 + 14r - 24$.

a) $(5r - 8)(r - 6)$　　　　b) $(5r - 4)(r + 6)$

c) $(5r - 12)(r + 2)$　　　　d) $(5r - 6)(r + 4)$

11. _____

12. _____

13. _____

14. _____

15. _____

16. _____

17. _____

18. Factor: $x^2 - 4xy + 4y^2 - 9z^2$.

     a) $(x - 2y + 3z)(x - 2y + 3z)$

     b) $(x - 2y - 3z)(x - 2y + 3z)$

     c) $(x + 2y - 3z)(x + 2y + 3z)$

     d) $x(x - 4y) + (2y - 3z)(2y + 3z)$

19. Write an expression containing a single radical.

$$\frac{\sqrt[5]{(x - 2)^2}\ \sqrt{x - 2}}{\sqrt[4]{(x - 2)^3}}$$

     a) $\sqrt[20]{(x - 2)^7}$ b) $\sqrt[20]{(x - 2)^3}$

     c) $\sqrt[10]{(x - 2)^3}$ d) $\sqrt[10]{(x - 2)^7}$

20. Convert to a radical notation: $y^{5/8}$.

     a) $\left(\sqrt{y^8}\right)^5$ b) $\left(\sqrt[5]{y}\right)^8$ c) $\sqrt[8]{5y}$ d) $\sqrt[8]{y^5}$

21. Factor and simplify: $7a^{2/3}b^{-1/2} - 4a^{-1/3}b^{1/2}$.

     a) $\dfrac{7a - 4b}{a^{1/3}b^{1/2}}$ b) $\dfrac{7a^{1/3} - 4b^{1/2}}{a^{1/3}b^{1/2}}$

     c) $\dfrac{7a - 4b}{a^{2/3}b^{1/2}}$ d) $\dfrac{7a - 4b}{a^{1/3}b^2}$

22. How long is a guy wire reaching from the top of a 14-m pole to a point 9 m from the base of the pole?

     a) 10.72 m   b) 16.64 m   c) 196 m   d) 277 m

23. Change $120\ \dfrac{km}{hr}$ to $\dfrac{m}{min}$.

     a) 2000   b) 200   c) 2   d) $33\frac{1}{3}$

ANSWERS

18. _____

19. _____

20. _____

21. _____

22. _____

23. _____

ANSWERS	
24. _____	

24. Consider the numbers:

$$-\sqrt{80}, \ \pi, \ \frac{8}{5}, \ \sqrt{121}, \ 0, \ 0.\overline{14}, \ 5\frac{2}{3}, \ -7.7.$$

List all the irrational numbers.

a) $-\sqrt{80}, \ \sqrt{121}$ b) $0, \ 0.\overline{14}$

c) $-\sqrt{80}, \ \pi$ d) $-\sqrt{80}, \ 0.\overline{14}, \ \pi$

25. _____

25. Multiply and simplify: $\dfrac{x^2 - 4}{x + 3} \cdot \dfrac{x^2 - 2x - 15}{x^2 - 3x - 10}.$

a) $x - 2$ b) 2

c) $\dfrac{x^2 + x - 6}{x + 3}$ d) $x + 2$

26. _____

26. Add and simplify: $\dfrac{9 + x}{x - 2} + \dfrac{5}{2 - x}.$

a) $\dfrac{14 + x}{x - 2}$ b) $\dfrac{5 - x}{x - 2}$ c) $\dfrac{14 + x}{2 - x}$ d) $\dfrac{x + 4}{x - 2}$

27. Divide and simplify: $\dfrac{x^2 + 3x - 28}{x^2 + 6x + 9} \div \dfrac{x - 4}{x + 3}.$

27. _____

a) $\dfrac{(x + 7)(x - 1)}{(x + 3)(x + 2)}$ b) $\dfrac{x + 7}{x + 3}$

c) $\dfrac{x - 4}{x + 3}$ d) $\dfrac{x}{x + 3}$

28. Subtract and simplify: $\dfrac{1}{x^2 - 16} - \dfrac{7}{x^2 - 2x - 8}.$

28. _____

a) $\dfrac{x - 5}{(x - 4)(x + 2)(x + 4)}$ b) $\dfrac{-2}{(x - 4)^2(x + 2)}$

c) $\dfrac{-2(3x + 13)}{(x - 4)(x + 2)(x + 4)}$ d) $\dfrac{-7x - 27}{(x - 4)(x + 2)(x + 4)}$

29. _____

29. Rationalize the denominator: $\dfrac{5x}{7 - \sqrt{x}}.$

a) $\dfrac{5(7 - \sqrt{x})}{-1}$ b) $\dfrac{35x + 5}{7 - x}$

c) $\dfrac{35x + 5x\sqrt{x}}{49 - x}$ d) $\dfrac{35x + 5x\sqrt{x}}{49 + x}$

NAME _____

CLASS _____ SCORE _____ GRADE _____

Simplify.

ANSWERS

1. i^{33} 2. $\sqrt{-7}\,\sqrt{-5}$ 3. $(4 - i)(2 + 5i)$

4. $(1 + 7i) - (-2 - 3i)$ 5. $\dfrac{4 + 5i}{2 - 3i}$

6. $(5 - i)(5 + i)$

7. Find the reciprocal of $2 - i$ and express it in the form $a + bi$.

8. Solve for x and y: $3x + 2 - 5i = 14 + (2 - y)i$.

Solve.

9. $(3y - 5)(y + 7)(y - 1) = 0$ 10. $x^4 - 3x^2 - 4 = 0$

11. $3a^2 - 13a + 12 = 0$ 12. $4t^2 - 2t - 7 = 0$

13. $\dfrac{20}{x + 3} = \dfrac{8}{x}$ 14. $4x^2 - x + 5 = 0$

15. $\sqrt{y - 3} = \sqrt{y + 18} - 3$ 16. $\dfrac{9}{x - 2} - \dfrac{4}{x - 1} = -1$

17. $(a - 4)(a - 3) - 30 = 0$ 18. $18 - 5y < 28$

19. $x^3 + x^2 - 4x - 4 = 0$

20. $5x - 7(x - 6) = 2[x - 3(2x - 7)]$

1. _____

2. _____

3. _____

4. _____

5. _____

6. _____

7. _____

8. _____

9. _____

10. _____

11. _____

12. _____

13. _____

14. _____

15. _____

16. _____

17. _____

18. _____

19. _____

20. _____

NAME _____

ANSWERS	
21. _____	21. The length of a rectangle is three times the width. The perimeter is 56 m. Find the dimensions.
22. _____	22. The speed of a boat in still water is 10 mph. It travels 15 mi upstream and 15 mi downstream in a total time of 4 hr. What is the speed of the current?
23. _____	23. Solve by completing the square. Show your work. $2x^2 - 12x - 6 = 0$
24. _____	24. Determine the nature of the solutions of $x^2 - 4x + 4 = 0$ by evaluating the discriminant.
25. _____	25. Write a quadratic equation whose solutions are $\sqrt{2}$ and $-3\sqrt{2}$.
26. _____	26. Solve $F = \dfrac{km_1 m_2}{d^2}$ for m_1.
27. _____	27. A polygon has 9 diagonals. How many sides does it have?
28. _____	28. Find an equation of variation in which y varies jointly as x and z and inversely as the square of w, and $y = 123$ when $x = 3$, $z = 2$, and $w = 4$.
29. _____	29. The volume of wood V in a tree trunk varies jointly as the height h and the square of the girth g. If the volume is 2160 ft^3 when the height is 80 ft and the girth is 6 ft, what is the height when the volume is 2400 ft^3 and the girth is 8 ft?
30. _____	30. Determine whether these equations are equivalent: $4x - 2 = 17$ $3x + 1 = 16 - x$.
31. _____	31. Determine the meaningful replacements in the radical expression $\sqrt{15 - 3x}$.
32. _____	Solve.
33. _____	32. $x = \dfrac{2}{3 + x}$ 33. $\sqrt{6x} - \sqrt{2x} = 4$

Simplify.   **ANSWERS**

1. i^{71} 2. $\sqrt{-15}\ \sqrt{-2}$ 3. $(3 - 4i)(1 + i)$ 1. _____

2. _____

4. $(9 + 3i) + (2 - i)$ 5. $\dfrac{4 - 3i}{2 + i}$

3. _____

6. $(7 + i)(7 - i)$ 4. _____

5. _____

7. Find the reciprocal of $5 + 2i$ and express it in the
   form $a + bi$.   6. _____

7. _____

8. Solve for x and y: $-1 + (x - y)i = x + 3y + 7i$.

8. _____

Solve.
9. _____

9. $(y - 9)(y + 3)(3y - 1) = 0$

10. _____

10. $(x^2 - x)^2 - 5(x^2 - x) + 6 = 0$

11. _____

11. $2b^2 - 3b - 20 = 0$ 12. $5t^2 - 2t - 4 = 0$ 12. _____

13. _____

13. $\dfrac{8}{2x - 3} = \dfrac{16}{2 - 4x}$ 14. $2x^2 - 5x + 9 = 0$

14. _____

15. _____

15. $\sqrt{31 - y} + \sqrt{y + 1} = 8$ 16. $\dfrac{5x - 1}{3} - \dfrac{x + 2}{4} = 2$

16. _____

17. $(a + 3)(a - 1) - 12 = 0$ 18. $24 - 7x > 38$ 17. _____
         a

18. _____

19. $x^3 - 2x^2 - 9x + 18 = 0$

19. _____

20. $x - 2(x - 10) = 5[x - 2(x - 2)]$

20. _____

NAME _____

ANSWERS	
21. _____	21. Suppose that $1500 is invested at 8%, compounded annually. What amount will be in the account at the end of 5 years?
22. _____	22. A can do a certain job in 5 hr, B can do the same job in 6 hr, and C can do the same job in 10 hr. How long would the job take with all three working together?
23. _____	23. Solve by completing the square. Show your work. $5x^2 - 20x - 10 = 0$
24. _____	24. Determine the nature of the solutions of $2x^2 - x - 4 = 0$ by evaluating the discriminant.
25. _____	25. Write a quadratic equation whose solutions are −2 and 2.
26. _____	26. Solve $C = 2\pi r$ for π.
27. _____	27. The hypotenuse of a triangle is 12 ft longer than one leg. The other leg of the triangle measures 36 ft. Find the length of the hypotenuse.
28. _____	28. Find an equation of variation in which y varies jointly as x and z and inversely as the square of w, and y = 20 when $x = \frac{1}{2}$, z = 4, and w = 5.
29. _____	29. The length ℓ of rectangles of fixed area is inversely proportional to the width w. Suppose that the length is 80 cm when the width is 5 cm. Find the length when the width is 8 cm.
30. _____	30. Determine whether these equations are equivalent: $7x - 3 = 4$ $9x + 1 = 2x + 8.$
31. _____	31. Determine the meaningful replacements in the radical expression $\sqrt{x - 4}$.
32. _____	32. A car is driven 280 miles. If it had gone 14 mph faster, it could have made the trip in 1 hr less time. What was the original speed?
33. _____	33. Solve for x: $kx^2 + (3k - 2)x - 6 = 0.$

NAME _____

CLASS _____ SCORE _____ GRADE _____

Simplify.

1. i^{19}　　　　　2. $-\sqrt{-5}\,\sqrt{-3}$　　　　3. $(5 - 2i)(4 + i)$

4. $(10 + 2i) - (-4 - 5i)$　　　5. $\dfrac{6 - i}{4 + 3i}$

6. $(3 - i)(3 + i)$

7. Find the reciprocal of $6 - 5i$ and express it in the form $a + bi$.

8. Solve for x and y:　$-3 + (x + y)i = 3x - 9y + 7i$.

Solve.

9. $(5y - 2)(y + 4)(y - 5) = 0$　　10. $x - 2\sqrt{x} - 8 = 0$

11. $5a^2 - 13a - 6 = 0$　　　　12. $3x^2 - 4x - 2 = 0$

13. $\dfrac{10}{3x - 4} = \dfrac{5}{x + 3}$　　　　14. $2x^2 + x + 4 = 0$

15. $\sqrt{y + 15} - \sqrt{14 - y} = 3$　　16. $\dfrac{2x + 4}{7} + \dfrac{x - 1}{4} = 3$

17. $(a + 2)(a - 5) - 8 = 0$　　18. $29 - 3x < 17$

19. $x^3 + 3x^2 - x - 3 = 0$

20. $3x - 5(x + 6) = 3[x - 5(2 - x)]$

ANSWERS

1. _____
2. _____
3. _____
4. _____
5. _____
6. _____
7. _____
8. _____
9. _____
10. _____
11. _____
12. _____
13. _____
14. _____
15. _____
16. _____
17. _____
18. _____
19. _____
20. _____

NAME _____

ANSWERS	
21. _____	21. In triangle ABC, angle A is twice as large as angle B. Angle C measures $20°$ less than angle B. Find the measures of the angles.
22. _____	22. A boat travels 215 km downstream in the same time that it takes to travel 150 km upstream. The speed of the current in the stream is 6.5 km/h. Find the speed of the boat in still water.
23. _____	23. Solve by completing the square. Show your work. $2x^2 - 8x - 4 = 0$
24. _____	24. Determine the nature of the solutions of $3x^2 - x + 4 = 0$ by evaluating the discriminant.
25. _____	25. Write a quadratic equation whose solutions are 4 and $-\dfrac{3}{2}$.
26. _____	26. Solve $P = 2\ell + 2w$ for ℓ.
27. _____	27. The area of a triangle is 42 cm^2. The base is 5 cm longer than the height. Find the height.
28. _____	28. Find an equation of variation in which y varies jointly as x and z, and y = 36 when x = 2 and z = 4.
29. _____	29. The distance s that an object falls from some point above the ground varies directly as the square of the time t that it falls. If the object falls 36 ft in 1.5 sec, how long will it take the object to fall 144 ft?
30. _____	30. Determine whether these equation are equivalent: $3x + 7 = 2$ $x + 5 = 2x.$
31. _____	31. Determine the meaningful replacements in the radical expression $\sqrt{5x - 1}$.
32. _____	Solve. 32. $3y^2 - (2y - 3)(y + 1) = 5$
33. _____	33. $(x - 6)^{2/3} = 3$

NAME _____

CLASS _____ SCORE _____ GRADE _____

Simplify.

1. i^{45} 2. $-\sqrt{-10}\ \sqrt{-3}$ 3. $(6 - i)(2 + 3i)$

4. $(-5 - 2i) + (7 - 3i)$ 5. $\dfrac{5 - 2i}{3 - 4i}$

6. $(4 - i)(4 + i)$

7. Find the reciprocal of $7 + 3i$ and express it in the form $a + bi$.

8. Solve for x and y: $5x + 3 + 2i = 8 + (3 + y)i$.

Solve.

9. $(2y - 3)(y - 4)(y - 1) = 0$ 10. $y^{2/3} - 2y^{1/3} - 3 = 0$

$(2x \quad 9)(x \quad 1)$

11. $7b^2 + 26b - 8 = 0$ 12. $2x^2 - 5x - 4 = 0$

$(2x - 2)(x + 2)$

13. $\dfrac{15}{x + 3} = \dfrac{6}{x}$ 14. $3x^2 - 2x + 3 = 0$

$(3x \quad 3)(x \quad)$

15. $\sqrt{y - 2} + 1 = \sqrt{y + 9}$ 16. $\dfrac{6}{2x - 1} + \dfrac{3}{x + 4} = 1$

17. $(a + 4)(a - 2) - 16 = 0$ 18. $5 - 9x > -13$

19. $x^3 + x^2 - 16x - 16 = 0$

20. $5x - 7(x - 4) = 2[x - 7(x - 2)]$

ANSWERS

1. _____

2. _____

3. _____

4. _____

5. _____

6. _____

7. _____

8. _____

9. _____

10. _____

11. _____

12. _____

13. _____

14. _____

15. _____

16. _____

17. _____

18. _____

19. _____

20. _____

ANSWERS	
21. _____	21. An investment is made at 6% compounded annually. It grows to $1404.50 at the end of one year. How much was originally invested?
22. _____	22. Bill can mow a lawn twice as fast as Sam. Working together it takes them 2 hours to mow the lawn. How long would it take each of them, working alone, to do the job?
23. _____	23. Solve by completing the square. Show your work. $3x^2 - 12x - 9 = 0$
24. _____	24. Determine the nature of the solutions of $4x^2 - 5x - 3 = 0$ by evaluating the discriminant.
25. _____	25. Write a quadratic equation whose solutions are 3i and -3i.
26. _____	26. Solve $\dfrac{P_1 V_1}{T_1} = \dfrac{P_2 V_2}{T_2}$ for V_1.
27. _____	27. The hypotenuse of a triangle is 39 ft. One leg is 6 ft longer than twice the other. What are the lengths of the legs?
28. _____	28. Find an equation of variation in which y varies jointly as x and z and inversely as the product of w and p, and $y = \dfrac{5}{8}$ when x = 2, z = 4, w = 5, and p = 9.
29. _____	29. The time t required to drive a fixed distance varies inversely as the speed r. It takes 6 hr at 55 mph to drive a fixed distance. How long would it take to drive the fixed distance at 40 mph?
30. _____	30. Determine whether these equations are equivalent: $5x - 9 = -2$ $3x + 5 = 12 - 2x.$
31. _____	31. Determine the meaningful replacements in the radical expression $\sqrt{3x - 4}$.
32. _____	32. For interest compounded annually, what is the interest rate when $7550 grows to $9580 in 3 years?
33. _____	33. Solve: $\sqrt{x - 5} - \sqrt[4]{x - 5} = 6$.

NAME _____

CLASS _____ SCORE _____ GRADE _____

Matching. In Questions 1 - 5, simplify. In Questions 6 - 15, solve. Place the letter of the answer in the answer blank. Some letters may be used more than once, and some may not be used.

1. $\left(\sqrt{3} - 4i\right)\left(\sqrt{3} + 4i\right)$

2. $\dfrac{i}{5 - i}$

3. $\dfrac{3 + 3i}{(3 - 3i)^2}$

4. $(9 - 27i) - (2 - 19i)$

5. $i^{20} - i^{43} + i^{17} - i^{105}$

6. $5t^2 - 3t + 2 = 0$

7. $\sqrt{x + 25} = \sqrt{40 - x} + 3$

8. $x^4 - 13x^2 + 36 = 0$

9. $\dfrac{5x}{x + 3} + \dfrac{15}{x} = \dfrac{45}{x^2 + 3x}$

10. $6z^2 - 10z - 5 = 0$

11. $-7x + 5 > -9$

12. $\dfrac{7}{x + 1} = \dfrac{3}{x + 3}$

13. $(a + 4)(a + 1) - 18 = 0$

14. $(y^2 - 5) + 2y = 5y - 1$

15. $4[2(x - 3) - 4(x + 2)] = 4x - 2(x - 2)$

ANSWERS

a) $-\dfrac{1}{6} + \dfrac{1}{6}i$

b) $x < -2$

c) $-7, 2$

d) $11 + 46i$

e) $7 - 8i$

f) $-\dfrac{1}{4} - \dfrac{1}{4}i$

g) $\dfrac{2 \pm 2i\sqrt{3}}{3}$

h) 19

i) $1 + i$

j) $2, 4, -4$

k) $2, -2, 4, -4$

ℓ) $-\dfrac{2}{5}, 1$

m) $0, -3$

n) $\dfrac{3 \pm i\sqrt{31}}{10}$

o) $-4\dfrac{1}{2}$

p) $-\dfrac{1}{26} + \dfrac{5}{26}i$

q) $-1, 4$

r) 24

s) $x < 2$

t) $\dfrac{5 \pm \sqrt{55}}{6}$

u) $24, -9$

v) -6

w) $-2, 2, -3, 3$

x) -13

y) 0

z) $\varnothing$

ANSWERS

1. _____

2. _____

3. _____

4. _____

5. _____

6. _____

7. _____

8. _____

9. _____

10. _____

11. _____

12. _____

13. _____

14. _____

15. _____

NAME _____

16. _____

17. _____

18. _____

19. _____

20. _____

21. _____

22. _____

23. _____

24. _____

25. _____

26. _____

27. _____

28. _____

29. _____

30. _____

16. Solve for x and y: $6 - (3 - y)i = x - 1 + 7i$.

17. Ship A can fill an oil storage tank 3 times as fast as ship B. When working together they can fill the oil storage tank in 6 hours. How long would it take ship A to fill the oil storage tank alone?

18. A boat can move at a speed of 20 km/h in still water. It will move 84 km downstream in a river in the same time it takes to move 36 km upstream. What is the speed of the river?

19. An investment is made at 11% compounded annually. It grows to $3885 at the end of one year. How much was originally invested?

20. Solve by completing the square. Show your work.

$4x^2 - 8x + 1 = 0$

21. Determine the nature of the solutions of

$5z^2 - 7z + 2 = 0$ by evaluating the discriminant.

22. Without solving, find the sum and the product of the solutions of $x^2 - 7x + 3 = 0$.

23. Write a quadratic equations whose solutions are $1 - i$ and $1 + i$.

24. Solve $a = \sqrt{2b - c}$ for b.   25. Solve $\frac{1}{a} = \frac{1}{b} + \frac{1}{c}$ for c.

26. Find the length of the side of a square inscribed in a circle of diameter 12.

27. Find an equation of variation where y varies directly as the cube of x and inversely as the sum of s and w, and $y = \frac{2}{5}$ when $x = 2$, $s = 1$, and $w = 3$.

28. In a circuit with constant voltage the current varies inversely as the resistance of the circuit. In a specially designed circuit the current is 7 amps when the resistance is 4 ohms. How many amps are there in the circuit when the resistance is 6 ohms?

29. Determine the meaningful replacements in the radical expression $\sqrt{5 + 12x}$.

30. Determine whether these equations are equivalent:

$\frac{y^2 - 9}{y + 3} = y - 3$ $y - 3 = y - 3$.

NAME _____

CLASS _____ SCORE _____ GRADE _____

Write the letter of your response on the answer blank.

ANSWERS

1. Simplify: $1^{27} - i^{15}$.

   a) $-i$ b) -1 c) $2i$ d) 0

1. _____

2. Simplify: $(4 - i)(-2 + 3i)$.

   a) $-6 + 3i$ b) $-5 + 14i$ c) 9 d) $-11 + 14i$

2. _____

3. Simplify: $(2 - 4i) - (-2 - i)$.

   a) $-5i$ b) $4 - 5i$ c) $6i$ d) $4 - 3i$

3. _____

4. Simplify: $\dfrac{3 + i}{2 - 4i}$.

   a) $\dfrac{1}{10} + \dfrac{7}{10}i$ b) $\dfrac{1}{6} + \dfrac{7}{6}i$ c) $\dfrac{1}{10} + \dfrac{1}{2}i$ d) $\dfrac{1}{3} - \dfrac{2}{3}i$

4. _____

5. Solve for x and y: $3x - 4 - 6i = 12 + (5 + y)i$.

   a) $x = \dfrac{8}{3}$, $y = -1$ b) $x = 2$, $y = -11$

   c) $x = 3$, $y = -1$ d) $x = \dfrac{16}{3}$, $y = -11$

5. _____

6. Solve: $10x^2 + 3 = 17x$. Find the larger solution.

   a) $\dfrac{1}{5}$ b) $\dfrac{2}{3}$ c) $\dfrac{3}{2}$ d) 3

6. _____

7. Solve: $\dfrac{4}{x - 3} - \dfrac{9}{x + 1} = 0$.

   a) $\dfrac{5}{8}, 3$, or $\dfrac{22}{5}$ b) $5, -\dfrac{31}{5}$, or $-\dfrac{5}{8}$

   c) $-1, \dfrac{31}{5}$, or 7 d) There are no solutions.

7. _____

NAME _____

ANSWERS

8. _____

9. _____

10. _____

11. _____

12. _____

13. _____

14. _____

8. Solve: $y(3y + 1)(y - 7) = 0$.

a) $-1, 0, 5$ b) $-\frac{1}{3}, 7$ c) $0, 7$ d) $-\frac{1}{3}, 0, 7$

9. Solve: $4x^2 + 1 = -4x$.

a) There is no solution.

b) There is just one solution, and it is negative.

c) There are two solutions, one positive
and one negative.

d) There are two solutions, both positive.

10. Solve: $2x - 5(x - 1) = 2[x - 3(x + 2)]$.

a) -1 b) 7 c) -17 d) $\frac{17}{7}$

11. Solve: $-3x + 7 < -14$.

a) $x > 7$ b) $x < -7$ c) $x > -7$ d) $x > \frac{7}{3}$

12. Solve and describe the solutions:

$5(y - 4) = (y - 4)(y - 1)$.

a) There are two solutions, both positive.

b) There are two solutions, one positive and one
negative.

c) There is just one solution.

d) There is no solution.

13. Solve: $4x^2 - 6x + 3 = 0$.

a) $\frac{3 \pm 2i\sqrt{3}}{4}$ b) $\frac{-3 \pm i\sqrt{5}}{4}$

c) $\frac{-2 \pm i\sqrt{3}}{4}$ d) $\frac{3 \pm i\sqrt{3}}{4}$

14. Solve: $\sqrt{3x - 2} - \sqrt{x - 5} = 3$.

a) The answer is one odd integer.

b) The answer is one negative integer.

c) The answer is two odd integers.

d) The answer is one even integer and one odd integer.

---

15. Grain flows through spout A twice as fast as through spout B. When grain flows through both spouts, a grain bin is filled in 6 hours. How many hours would it take to fill the grain bin if grain flows through spout B alone?

    a) 4   b) 18   c) 8   d) 16

16. The speed of car A is 15 mph faster than the speed of car B. Car A travels 300 mi in the same time it takes car B to travel 225 mi. Find the speed of car A.

    a) 60 mph   b) 45 mph   c) 50 mph   d) 65 mph

17. Determine the nature of the solutions of

    $3x^2 - 7x + 10 = 0$ by evaluating the discriminant.

    a) Two real solutions   b) One real solution

    c) Two non-real solutions   d) One integer solution

18. Find a quadratic equation having the solutions $2 - 5i$, $2 + 5i$.

    a) $x^2 - 4x + 29 = 0$ b) $x^2 + 4 = 0$

    c) $x^2 + 4x + 29 = 0$ d) $x^2 + 25 = 0$

19. Solve for x: $C = \sqrt{2 - \dfrac{x^2}{y^2}}$.

    a) $x = \dfrac{\sqrt{C^2 - y^2}}{2}$ b) $x = y\sqrt{2 - C^2}$

    c) $x = \sqrt{y^2 + C^2 y^2}$ d) $x = y\sqrt{C^2 - 2}$

ANSWERS

15. _____

16. _____

17. _____

18. _____

19. _____

NAME _____

20. _____

21. _____

22. _____

23. _____

24. _____

20. Solve: $\dfrac{7}{x} - \dfrac{3}{x^2} = 2$. Find the sum of the roots.

 a) $2\frac{1}{2}$ b) $1\frac{1}{2}$ c) $-3\frac{1}{2}$ d) $3\frac{1}{2}$

21. A picture frame measures 10 cm by 18 cm; 48 cm^2 of picture shows. Which of the following equations can be used to find the width of the frame?

 a) $2x + 28 = 48$ b) $(18 + 2x)(10 + 2x) = 48$

 c) $x^2 - 14x + 33 = 0$ d) $10x \cdot 18x = 48$

22. Find an equation of variation in which y varies inversely as the cube of x, and y = 5 when x = 2.

 a) $y = 10x^3$ b) $y = \dfrac{40}{x^3}$ c) $y = \dfrac{1.6}{x^3}$ d) $y = 0.625x^3$

23. The force of wind on a sail varies jointly as the area of the sail and the square of the wind velocity. On a square foot of a sail the force is 4 lb when the wind velocity is 20 mph. Find the force of a 16-mph wind on a sail of area 50 sq. ft.

 a) 128 lb b) 400 lb c) 320 lb d) 160 lb

24. Solve: $x^3 - x^2 = 1 - x$.

 a) There is just one solution.

 b) There are two solutions, both negative.

 c) There are three solutions.

 d) There are two solutions, one negative and one positive.

NAME _____

CLASS _____ SCORE _____ GRADE _____

Consider the relation {(2,8), (-1,-5), (8,7), (5,1)} for Questions 1 - 3.

1. Determine whether the relation is a function.

2. Find the range.

3. Find the domain.

Graph.

4. $f(x) = (x + 1)^2$ 5. $y = |x - 2|$ 6. $f(x) = 3\ \text{INT}(x)$

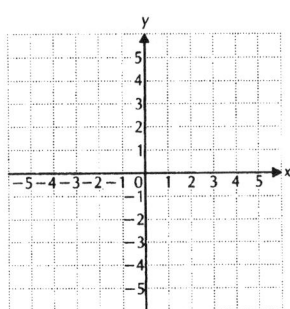 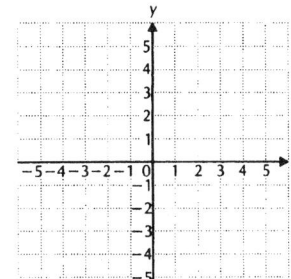 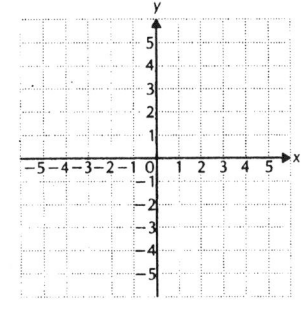

Consider the following relations for Questions 7 and 8.

a) $3x - \dfrac{2}{y} = 0$ b) $x^3 + 2y = y^2$ c) $x = -2$

d) $x^2 + y^2 = 1$ e) $y = x^2 - 4$ f) $4y = |2x|$

7. Which are symmetric with respect to the origin?

8. Which are symmetric with respect to the x-axis?

9. Find the domain of the function f given by $\sqrt{x - 2}$.

Consider the functions f and g given by $f(x) = 2x^2$ and $g(x) = 3x - 5$ for Questions 10 and 11.

10. Find $(f + g)(x)$, $(f - g)(x)$, $fg(x)$, $(f/g)(x)$, $f \circ g(x)$, and $g \circ f(x)$.

11. Find the domain of f, g, f + g, f - g, fg, f/g, f∘g, and g∘f.

12. Which of the following are graphs of functions?

a)

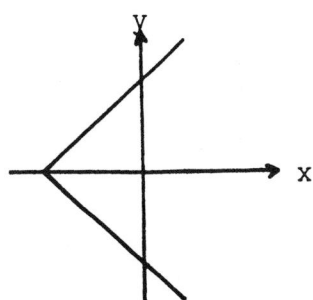

b)

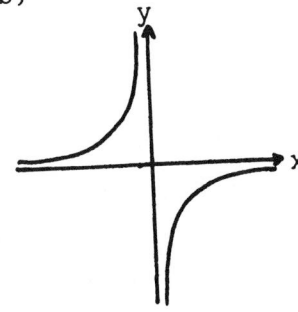

ANSWERS

1. _____

2. _____

3. _____

4. _See graph._

5. _See graph._

6. _See graph._

7. _____

8. _____

9. _____

10. _____

_____

_____

_____

_____

_____

_____

11. _____

_____

_____

_____

_____

_____

_____

12. _____

NAME _____

TEST FORM A

ANSWERS	

Use $g(x) = x^2 - 3x + 4$ for Questions 13 - 16. Find:

13. _____

13. $g(-2)$ 14. $g(0)$

14. _____

15. $g(a - 2)$ 16. $\dfrac{g(a + h) - g(a)}{h}$

17. Given $R(x) = 5x^2 - 17x - 42$ and $C(x) = 3.9x^2 + 15x - 12$, find $P(x)$.

15. _____

18. Find the slope and the y-intercept of the line $15 = 7x - 2y$.

16. _____

19. Find an equation of the line through $(-1,7)$ with $m = 2$.

20. Find an equation of the line containing $(4,-1)$ and $(6,2)$.

17. _____

21. Find the distance between $(3,-4)$ and $(7,-2)$.

22. Find the midpoint of the segment with endpoints $(3,-2)$ and $(6,1)$.

18. _____

23. Determine whether these lines are parallel, perpendicular, or neither.

$$5x - y = 11$$
$$x = \frac{y}{5} + 3$$

19. _____

20. _____

24. Graph $y = \frac{2}{3}x - 1$ using the slope and the y-intercept.

21. _____

22. _____

23. _____

24. See graph.

25. _____

25. Find an equation of the line containing the given point and perpendicular to the given line.

$$(3,-1); \ y = \frac{1}{4}x - 5$$

26. _____

26. Find an equation of the circle with center $(3,-4)$ and radius 2.

27. _____

27. Find the center and the radius of the circle

$$x^2 + y^2 - 8x + 2y + 1 = 0.$$

TEST FORM A

	ANSWERS

28. Here is a graph of y = f(x). Sketch the graph of each of the following.

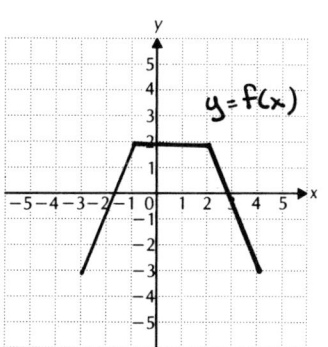

28. a) See graph.

    b) See graph.

    c) See graph.

a) y = f(x + 1)   b) y = f(2x)   c) y = 3 + f(x)

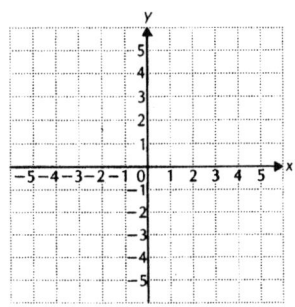 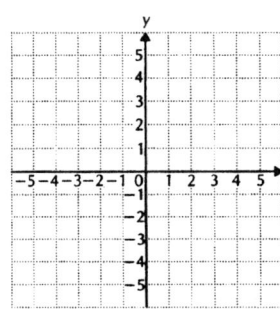 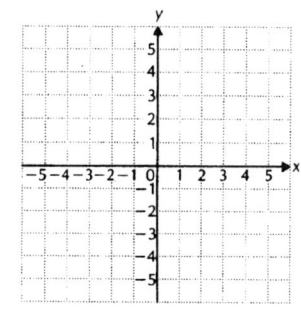

29. _____

Use the following for Questions 29 – 31.

a)

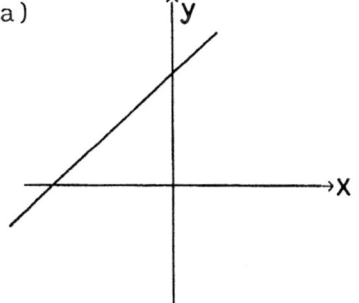

b)

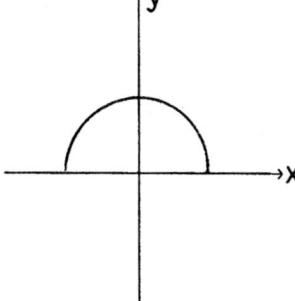

30. _____

c) f(x) = 3x^2 + 5x   d) f(x) = x^3 – 4x + 1

e) f(x) = $\sqrt{x}$ f) f(x) = x^9 – x^3

31. _____

29. Which are even?   30. Which are odd?

31. Which are neither even nor odd?

ANSWERS	

Write interval notation for the set.

32. _____

32. $\{x \mid -1 < x < 3\}$ 33. $\{x \mid x \geq -7\}$

Use the following for Questions 34 and 35.

33. _____

a) 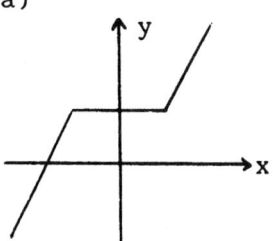   b) 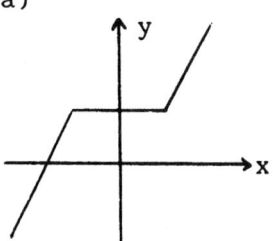   c)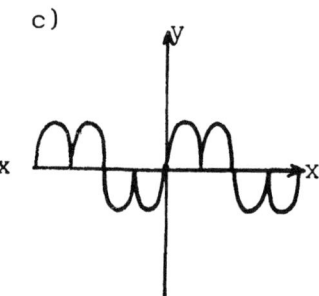

34. _____

35. _____

34. Which are decreasing?   35. Which are increasing?

36. Graph.

$$f(x) = \begin{cases} x^2, & \text{for } x < -1, \\ x, & \text{for } -1 \leq x < 3, \\ x + 2, & \text{for } x \geq 3 \end{cases}$$

36. See graph.

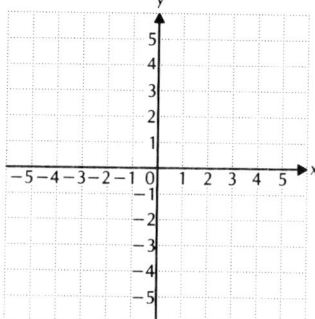

37. _____

37. A rectangle of dimensions x by y is inscribed in a circle of radius 6 ft. Express the area A as a function of x.

38. a) _____

b) _____

38. Find f(x) and g(x) such that h(x) = f∘g(x).

    a) $h(x) = 5|3x - 2|$

    b) $h(x) = 18 - 2(2x - 3)^2$

39. _____

39. Find the domain: $f(x) = (x - 4x^{-1})^{-1}$.

NAME _____

CLASS _____ *SCORE* _____ *GRADE* _____

Consider the relation {(7,1), (2,5), (-1,4), (7,-3)} for Questions 1 - 3.

1. Determine whether the relation is a function.

2. Find the range.

3. Find the domain.

Graph.

4. $f(x) = 2 - x^2$ 5. $x = |y - 3|$ 6. $f(x) = INT(x + 2)$

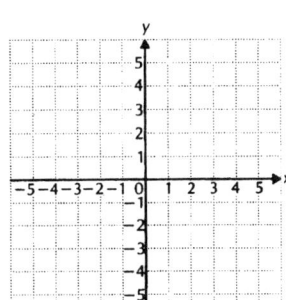 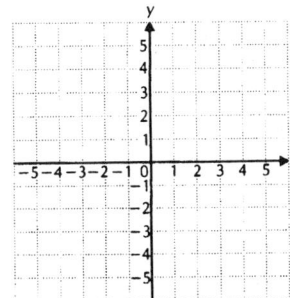 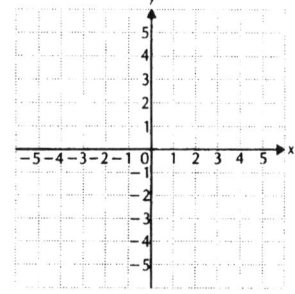

Consider the following relations for Questions 7 and 8.

a) $y = x^4 - 5$ b) $x^3 - x = y^2$ c) $x^2 - 3 = y$

d) $3x - \dfrac{5}{y} = 0$ e) $2x = |y|$ f) $x = -1$

7. Which are symmetric with respect to the origin?

8. Which are symmetric with respect to the x-axis?

9. Find the domain of the function f given by $\dfrac{1}{8 - x^3}$.

Consider the functions f and g given by $f(x) = 5x + 2$ and $g(x) = x^2 - 1$ for Questions 10 and 11.

10. Find $(f + g)(x)$, $(f - g)(x)$, $fg(x)$, $(f/g)(x)$, $f\circ g(x)$, and $g\circ f(x)$.

11. Find the domain of f, g, f + g, f - g, fg, f/g, f∘g, and g∘f.

12. Which of the following are graphs of functions?

a) 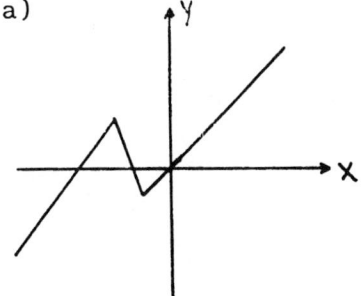   b)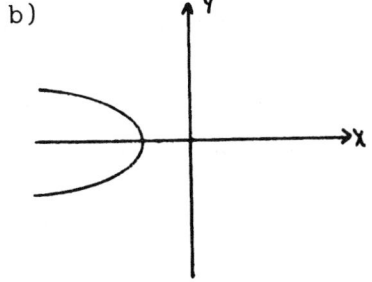

ANSWERS

1. _____

2. _____

3. _____

4. See graph. _____

5. See graph. _____

6. See graph. _____

7. _____

8. _____

9. _____

10. _____

_____

_____

_____

_____

_____

11. _____

_____

_____

_____

_____

_____

_____

_____

12. _____

TEST FORM B

ANSWERS	

Use $g(x) = 3x^2 + 8x$ for Questions 13 - 16. Find:

13. $g(-1)$ 14. $g(0)$

15. $g(a + 1)$ 16. $\dfrac{g(a + h) - g(a)}{h}$

17. Given $R(x) = 4x^2 + 45x + 200$ and $C(x) = 1.7x^2 + 8x - 90$, find $P(x)$.

18. Find the slope and the y-intercept of the line $20 - 3x = 2y$.

19. Find an equation of the line through $(-1,4)$ with $m = -2$.

20. Find an equation of the line containing $(-2,-3)$ and $(4,-2)$.

21. Find the distance between $(8,-3)$ and $(4,-2)$.

22. Find the midpoint of the segment with endpoints $(3,-7)$ and $(6,1)$.

23. Determine whether these lines are parallel, perpendicular, or neither.

$$3y - 2x = 7,$$
$$2y = 8 - 3x$$

24. Graph $y = \dfrac{3}{4}x + 2$ using the slope and the y-intercept.

25. Find an equation of the line containing the given point and parallel to the given line.

$$(-4,2); \quad y + 3 = 2x$$

26. Find an equation of the circle with center $(-1,-3)$ and radius 5.

27. Find the center and the radius of the circle $x^2 + y^2 - 4x + 10y + 4 = 0$.

ANSWERS

13. _____

14. _____

15. _____

16. _____

17. _____

18. _____

19. _____

20. _____

21. _____

22. _____

23. _____

24. See graph.

25. _____

26. _____

27. _____

28. Here is a graph of $y = f(x)$. Sketch the graph of each of the following.

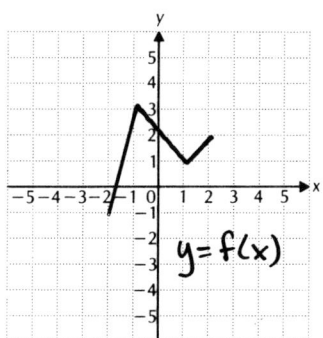

28. a) See graph.

b) See graph.

c) See graph.

a) $y = f(x) - 1$ b) $y = \frac{1}{2}f(x)$ c) $y = 3 + f(x)$

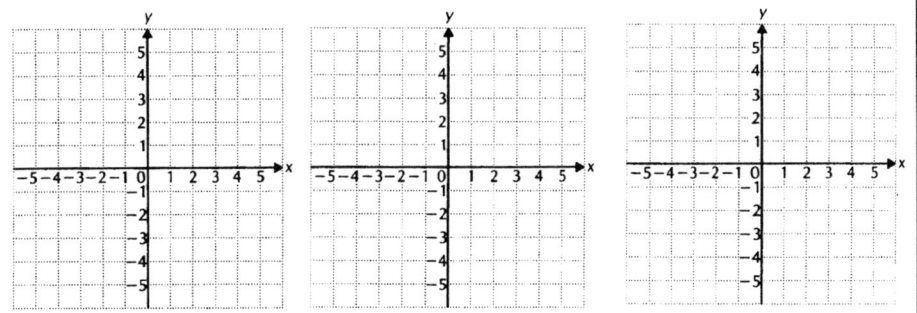

Use the following for Questions 29 - 31.

29. _____

a) 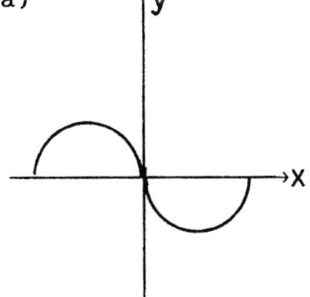   b)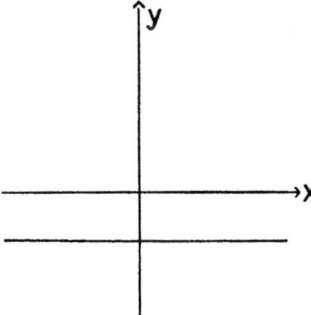

30. _____

c) $f(x) = \frac{1}{x}$ d) $f(x) = |2x|$

e) $f(x) = 4x^2 + 3x^4 - 2$ f) $f(x) = x^3 - 2x + 5$

31. _____

29. Which are even?   30. Which are odd?

31. Which are neither even nor odd?

NAME _____

ANSWERS	Write interval notation for the set.

32. _____

32. $\{x \mid x \le -1\}$ 33. $\{x \mid -9 < x \le 4\}$

Use the following for Questions 34 and 35.

33. _____

a) 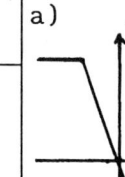   b)   c)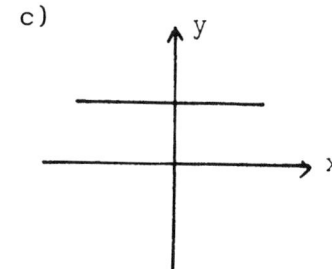

34. _____

35. _____

34. Which are increasing?   35. Which are neither increasing nor decreasing?

36. Graph.

36. See graph.

$$f(x) = \begin{cases} \dfrac{x}{2}, & \text{for } x \le 0, \\ x^2 - 1, & \text{for } 0 < x \le 2, \\ x + 3, & \text{for } x > 2 \end{cases}$$

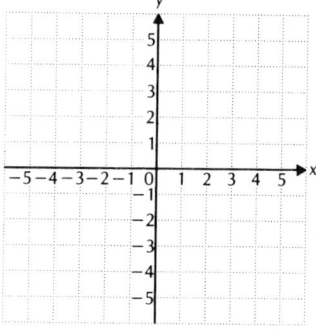

37. _____

37. The base b of a triangle is 4 less than twice the height h. Express the area A of the triangle as a function of the height h.

38. a) _____

38. Find f(x) and g(x) such that $h(x) = f \circ g(x)$.

   a) $h(x) = \dfrac{1}{(x + 3)^2}$

   b) _____

   b) $h(x) = 5(x - 3)^3 + 2$

39. _____

39. Find k so that the line containing $(-2, k)$ and $(4, 2)$ is parallel to the line containing $(7, 1)$ and $(2, -4)$.

Consider the relation {(3,5), (-1,-6), (5,3), (0,-1)} for Questions 1 - 3.

1. Determine whether the relation is a function.

2. Find the range.

3. Find the domain.

Graph.

4. $f(x) = x^2 - 5$ 5. $y = |x + 2|$ 6. $f(x) = 2\ \text{INT}(x - 1)$

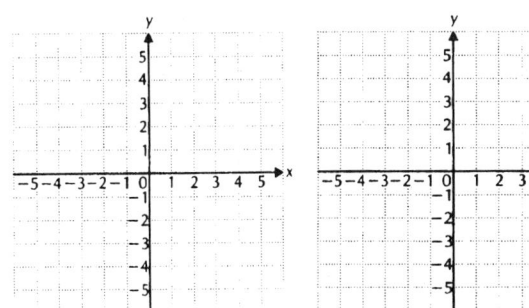
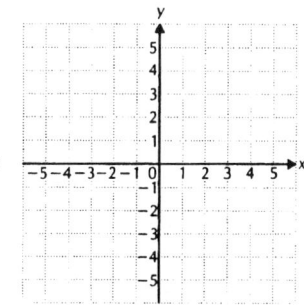

	ANSWERS
1.	_____
2.	_____
3.	_____
4.	See graph.
5.	See graph.
6.	See graph.
7.	_____
8.	_____
9.	_____
10.	_____

Consider the following relations for Questions 7 and 8.

a) $x^3 + 2 = y$ b) $\dfrac{2y}{3} - x = 4$ c) $x = |y|$

d) $y = 4$ e) $x = y^2 + 3$ f) $y^2 + x^2 = 5$

7. Which are symmetric with respect to the y-axis?

8. Which are symmetric with respect to the origin?

9. Find the domain of the function f given by $3 - \dfrac{5}{x}$.

Consider the functions f and g given by $f(x) = 9 - 4x$ and $g(x) = x^2 + 3$ for Questions 10 and 11.

11. _____

10. Find $(f + g)(x)$, $(f - g)(x)$, $fg(x)$, $(f/g)(x)$, $f \circ g(x)$, and $g \circ f(x)$.

11. Find the domain of f, g, f + g, f - g, fg, f/g, f∘g, and g∘f.

12. Which of the following are graphs of functions?

a)

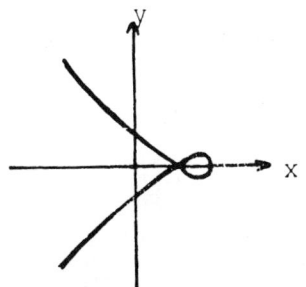

b)

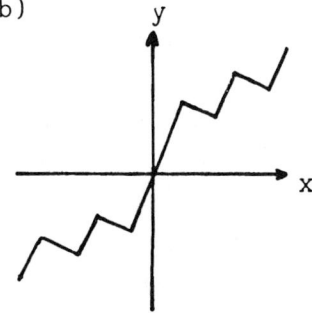

12. _____

ANSWERS	Use $g(x) = x^2 + 5x - 1$ for Questions 13 - 16. Find:

13. _____

14. _____

15. _____

16. _____

17. _____

18. _____

19. _____

20. _____

21. _____

22. _____

23. _____

24. See graph.

25. _____

26. _____

27. _____

13. $g(-3)$ 14. $g(0)$

15. $g(b + 2)$ 16. $\dfrac{g(a + h) - g(a)}{h}$

17. Given $R(x) = 2x^2 + 50x - 10$ and $C(x) = 1.5x^2 - 14x + 15$, find $P(x)$.

18. Find the slope and the y-intercept of the line $2x - 5y = 9$.

19. Find an equation of the line through $(0,-7)$ with $m = 2$.

20. Find an equation of the line containing $(4,-2)$ and $(-1,5)$.

21. Find the distance between $(3,4)$ and $(-4,-2)$.

22. Find the midpoint of the segment with endpoints $(-4,-2)$ and $(5,-3)$.

23. Determine whether these lines are parallel, perpendicular, or neither.

$$x - 3y = 7,$$
$$3x - y = 2.$$

24. Graph $y = -\dfrac{2}{3}x + 4$ using the slope and the y-intercept.

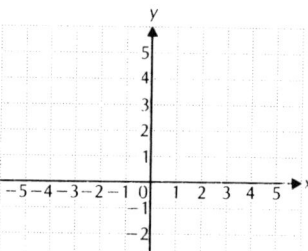

25. Find an equation of the line containing the given point and perpendicular to the given line.

$$(0,-5); \ 2y = x - 1$$

26. Find an equation of the circle with center $(-1,5)$ and radius 3.

27. Find the center and the radius of the circle $x^2 + y^2 - 6x + 4y + 8 = 0$.

28. Here is a graph of y = f(x). Sketch the graph of each of the following.

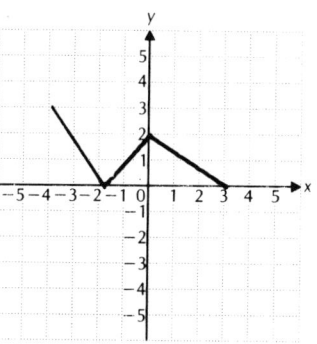

a) y = f(x - 1)   b) y = 2f(x)   c) y = 1 - f(x)

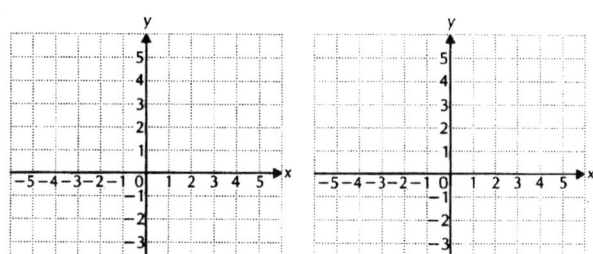

29. _____

Use the following for Questions 29 - 31.

a)

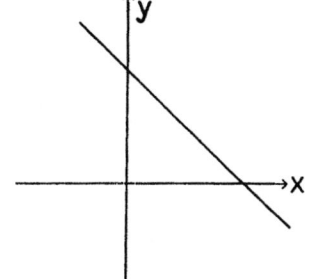

b)

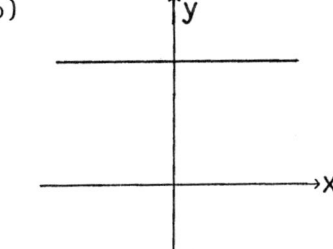

30. _____

c) f(x) = -2x^3 + 5x   d) f(x) = 3|x| - x

e) f(x) = x^9 f) f(x) = 5x^4 + 3x^2 - 6

31. _____

29. Which are even?   30. Which are odd?

31. Which are neither even nor odd?

NAME _____

ANSWERS	

Write interval notation for the set.

32. $\{x \mid -2 < x \le 5\}$ 33. $\{x \mid x < -4\}$

32. _____

Use the following for Questions 34 and 35.

a)   b)   c)

33. _____

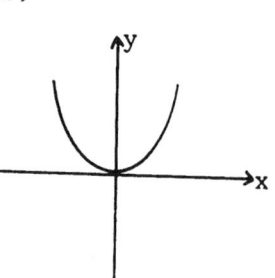   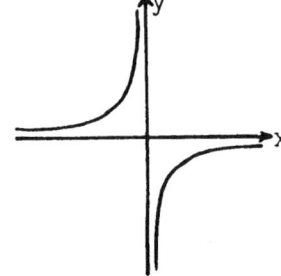

34. _____

35. _____

34. Which are decreasing?   35. Which are neither increasing nor decreasing?

36. Graph.

$$f(x) = \begin{cases} 4, & \text{for } x < -2, \\ x + 1, & \text{for } -2 \le x < 0, \\ x^2, & \text{for } x \ge 0 \end{cases}$$

36. See graph. _____

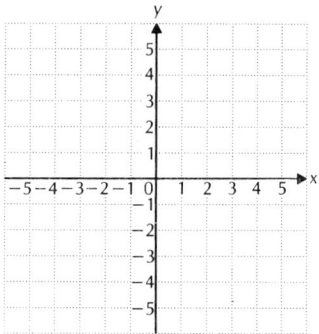

37. _____

37. A rectangle of dimensions x by y has a perimeter of 20 ft. Express the area A as a function of x.

38. a) _____

38. Find f(x) and g(x) such that $h(x) = f \circ g(x)$.

   a) $h(x) = \sqrt[3]{3x - 4}$

   b) _____

   b) $h(x) = 2(x - 4)^2 + (x - 4) + 7$.

39. _____

39. Find the point on the x-axis that is equidistant from the points (2,7) and (10,1).

NAME _____

CLASS _____ SCORE _____ GRADE _____

Consider the relation {(3,-1), (5,3), (-1,4), (3,5)}
for Questions 1 - 3.

1. Determine whether the relation is a function.

2. Find the range.

3. Find the domain.

Graph.

4. $f(x) = 4 - x^2$ 5. $x = |y + 1|$ 6. $f(x) = INT\ (x - 3)$

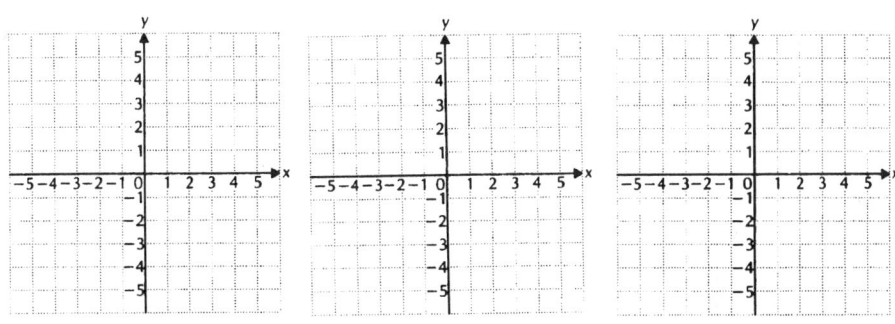

Consider the following relations for Questions 7 and 8.

a) $x^5 - 5 = y^2$ b) $\dfrac{3x}{2} - y = 5$ c) $x = 9$

d) $y = |x|$ e) $x = y^3 - 4$ f) $x^2 - y^3 = 2$

7. Which are symmetric with respect to the y-axis?

8. Which are symmetric with respect to the origin?

9. Find the domain of the function f given by
$$\sqrt{2x + 3}.$$

Consider the functions f and g given by $f(x) = 3x^2 - 1$ and
$g(x) = 2x + 4$ for Questions 10 and 11.

10. Find $(f + g)(x)$, $(f - g)(x)$, $fg(x)$, $(f/g)(x)$, $f \circ g(x)$,
and $g \circ f(x)$.

11. Find the domain of f, g, f + g, f - g, fg, f/g, f∘g,
and g∘f.

12. Which of the following are graphs of functions?

a) 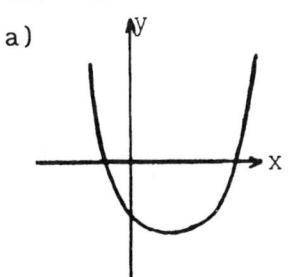   b)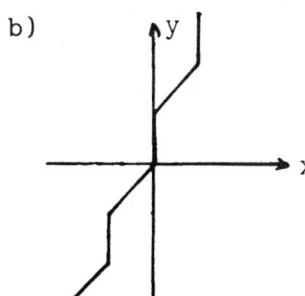

ANSWERS

1. _____

2. _____

3. _____

4. See graph.

5. See graph.

6. See graph.

7. _____

8. _____

9. _____

10. _____

_____

_____

_____

_____

_____

_____

_____

11. _____

_____

_____

_____

_____

_____

_____

_____

_____

12. _____

NAME _____

ANSWERS
13. _____
14. _____
15. _____
16. _____
17. _____
18. _____
19. _____
20. _____
21. _____
22. _____
23. _____
24. See graph.
25. _____
26. _____
27. _____

Use $g(x) = 2x^2 - 5x$ for Questions 13 - 16. Find:

13. $g(-4)$

14. $g(0)$

15. $g(a - 1)$

16. $\dfrac{g(a + h) - g(a)}{h}$

17. Given $R(x) = 2x^2 + 13x - 17$
$C(x) = 0.9x^2 - 5x + 12$, find $P(x)$.

18. Find the slope and the y-intercept of the line
$15 - 2x = 5y$.

19. Find an equation of the line through $(1,-2)$ with $m = -3$.

20. Find an equation of the line containing $(3,5)$ and
$(-4,1)$.

21. Find the distance between $(7,-2)$ and $(-1,3)$.

22. Find the midpoint of the segment with endpoints $(2,-4)$
and $(6,3)$.

23. Determine whether these lines are parallel,
perpendicular, or neither.

$$2x - y = 7,$$
$$5x = -10y + 3.$$

24. Graph $y = \dfrac{1}{2}x - 5$ using the slope and the y-intercept.

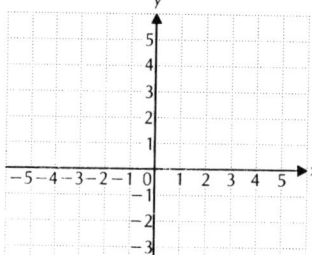

25. Find an equation of the line containing the given point
and parallel to the given line.

$$(1,-3); \ 2x = 4y - 3$$

26. Find an equation of the circle with center $(2,-6)$ and
radius 4.

27. Find the center and the radius of the circle
$x^2 + y^2 - 2x + 12y + 1 = 0$.

28. Here is a graph of y = f(x). Sketch the graph of each
    of the following.

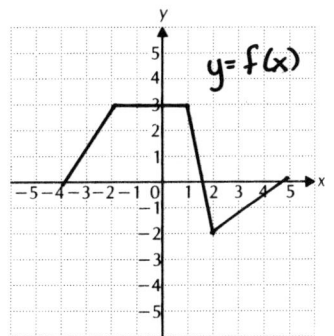

28. a) See graph.

    b) See graph.

    c) See graph.

a) $y = \frac{1}{2}f(x)$ b) $y = f(x) + 1$ c) $y = f(x - 2)$

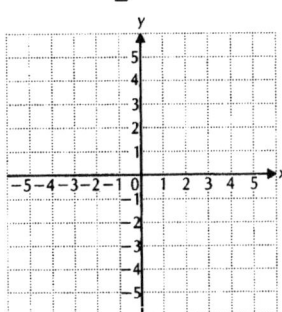   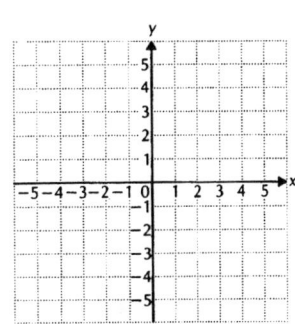   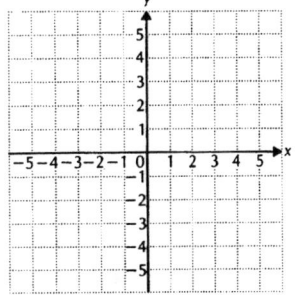

29. _____

Use the following for Questions 29 - 31.

a)

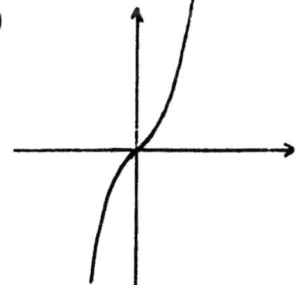

b)

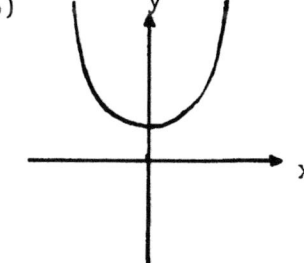

30. _____

c) $f(x) = \frac{3}{x}$ d) $f(x) = x^3 - 2x^2 + 1$

e) $f(x) = \sqrt{x + 2}$ f) $f(x) = 2x^4 - 3x^2$

31. _____

29. Which are even?   30. Which are odd?

31. Which are neither even nor odd?

51

NAME _____

ANSWERS	Write interval notation for the set.

32. _____

32. $\{x \mid -5 < x \le 2\}$ 33. $\{x \mid x \le 6\}$

Use the following for Questions 34 and 35.

a)   b)   c)

33. _____

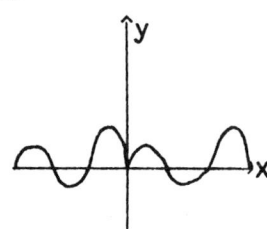   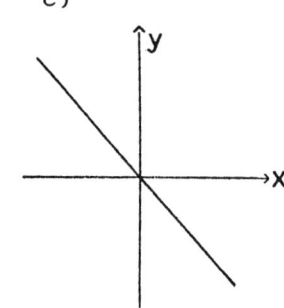

34. _____

35. _____

34. Which are decreasing?   35. Which are increasing?

36. Graph.

36. See graph.

$$f(x) = \begin{cases} x + 3, & \text{for } x < -3, \\ x, & \text{for } -3 \le x < 1, \\ x^2 + 1, & \text{for } x \ge 1 \end{cases}$$

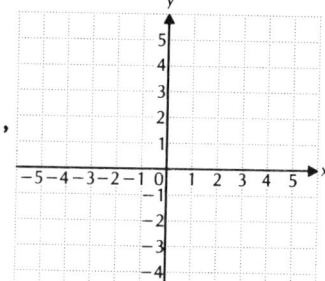

37. _____

37. A rectangle of dimensions x by y is inscribed in a circle of radius 10 ft. Express the area A as a function of x.

38. a) _____

38. Find f(x) and g(x) such that h(x) = f∘g(x).

   b) _____

a) $h(x) = \dfrac{1}{\sqrt{2x - 1}}$

b) $h(x) = 6(x + 2)^2 + 5$

39. _____

39. Find k so that the line containing (-1,k) and (3,1) is perpendicular to the line containing (6,2) and (-1,4).

NAME _____

CLASS _____ *SCORE* _____ *GRADE* _____

Consider the relation {(0,-1), (5,2), (-5,-1), (3,5), (-2,4)} for Questions 1 - 3.

1. Determine whether the relation is a function.

2. Find the domain.

3. Find the range.

Graph.

4. $x = |y + 1|$ 5. $y = x^2 + 2$ 6. $f(x) = INT\ (x - 1)$

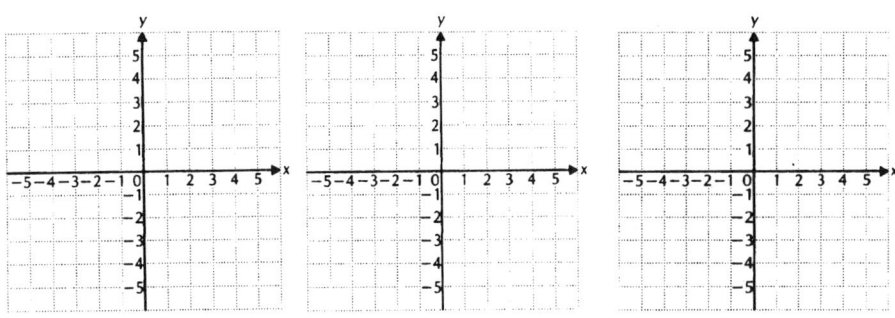

Consider the following relations for Questions 7 - 9.

a) $y = x^3$ b) $y = 3x^2 - 2x + 5$

c) $y = 7$ d) $x = \dfrac{2}{y}$

e) $4x - y = 6$ f) $4x^2 + 4y^2 = 8$

7. Which are symmetric with respect to the x-axis?

8. Which are symmetric with respect to the origin?

9. Which are symmetric with respect to the y-axis?

10. Find the domain: $f(x) = \dfrac{x - 4}{x^2 - 3x + 4}$.

Consider the functions f and g given by $f(x) = 2x - 3$ and $g(x) = 4x^2 - 1$ for Questions 11 and 12.

11. Find $(f + g)(x)$, $(f - g)(x)$, $fg(x)$, $(f/g)(x)$, $f \circ g(x)$, and $g \circ f(x)$.

12. Find the domain of f, g, f + g, f - g, fg, f/g, f ∘ g, and g ∘ f.

ANSWERS

1. _____

2. _____

3. _____

4. See graph.

5. See graph.

6. See graph.

7. _____

8. _____

9. _____

10. _____

11. _____

_____

_____

_____

_____

12. _____

_____

_____

_____

_____

_____

NAME

13. Which of the following are graphs of functions?

13. _____

a)

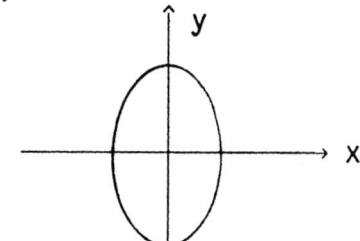

b)

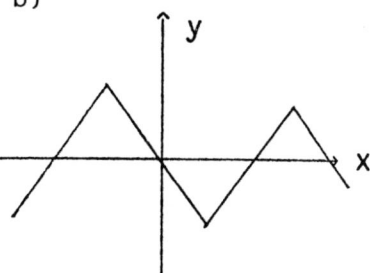

14. _____

c)

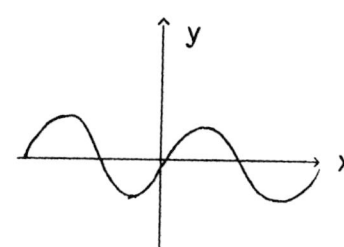

d)

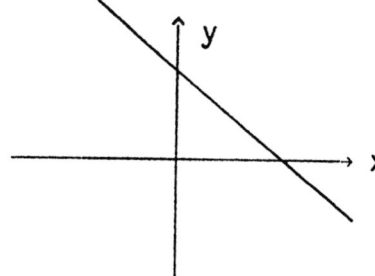

15. _____

16. _____

Use $g(x) = 3x^2 - x + 5$ for Questions 14 – 17. Find:

14. $g(4)$ 15. $g(-3)$

17. _____

16. $g(a - 2)$ 17. $\dfrac{g(a + h) - g(a)}{h}$

18. Given $R(x) = 30x - 0.2x^2$ and $C(x) = 3x - 2$, find $P(x)$.

18. _____

NAME _____

19. Here is a graph of y = f(x). Sketch the graph of each
    of the following.

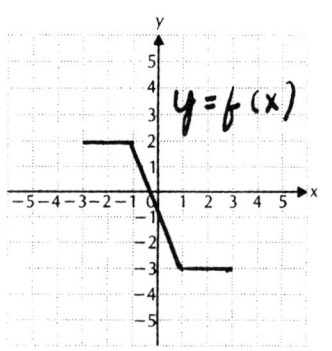

ANSWERS

19. a) See graph.

   b) See graph.

   c) See graph.

a) y = f(x + 1)   b) y = f(x) - 2   c) y = 2f(x)

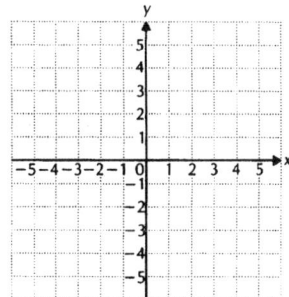 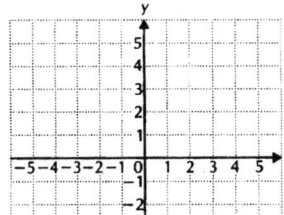 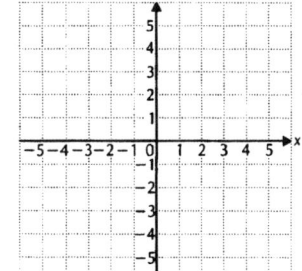

20. _____

Use the following for Questions 20 - 22.

a)   b)

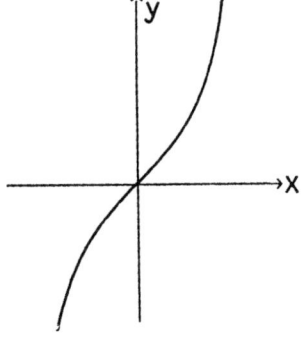 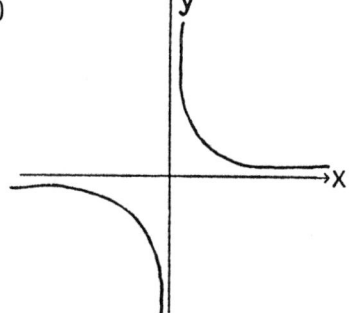

21. _____

22. _____

23. _____

c) f(x) = x^2 + 7   d) f(x) = x^3 - 2x - 8

e) f(x) = |3x| f) f(x) = 3x^5 - 2x

24. _____

20. Which are even?   21. Which are odd?

22. Which are neither even nor odd?

Write interval notation for the set.

23. {x|-2 ≤ x < 5}   24. {x|x > -3}

TEST FORM E

ANSWERS	

25. _____

26. _____

27. _____

28. _____

29. _____

30. See graph.

31. _____

32. _____

33. _____

34. _____

Use the following for Questions 25 and 26.

a) 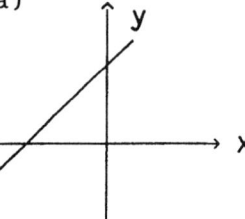 b) c)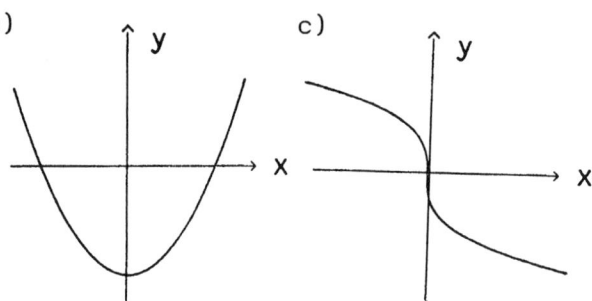

25. Which are increasing?

26. Which are neither increasing nor decreasing?

27. Find the slope and y-intercept of the line $5 = \frac{y}{2} + 3x$.

28. Find an equation of the line containing $(0,-7)$ and $(2,3)$.

29. Find the distance between $(7,-1)$ and $(6,2)$.

30. Graph.

$$f(x) = \begin{cases} x - 2 & \text{for } x \le -1, \\ |x| & \text{for } -1 < x \le 4, \\ \sqrt{x} & \text{for } x > 4 \end{cases}$$

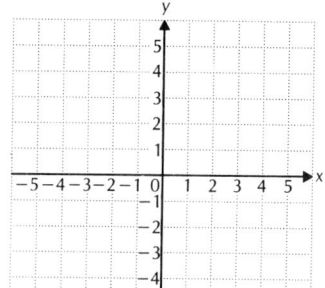

31. A rectangle of dimensions s by t is inscribed in a circle of radius 8 cm. Express the area A of the rectangle as a function of t.

32. Find an equation of the line containing the given point and perpendicular to the given line:

$$(-1,4); \quad x = \frac{2}{3}y - 1.$$

33. Find $f(x)$ and $g(x)$ such that $h(x) = f \circ g(x)$.

$$h(x) = \frac{5\sqrt{3x - 8}}{2}$$

34. Find an equation of the circle with center $(-1,5)$ and radius 12.

NAME _____

CLASS _____ SCORE _____ GRADE _____

Write the letter of your response on the answer blank.

ANSWERS

1. List the range of the relation whose ordered pairs are (2,4), (-2,4), (4,3), (0,-4), and (-2,3).

    a) {-4,0,3,4}   b) {-4,3,4}

    c) {-4,-4,0,3,4}   d) {2,2,3,4,4}

1. _____

Consider the following relations for Questions 2 - 4.

a) $y = x^5 - x^3$ b) $x^4 + y^4 = 2$ c) $\dfrac{y}{2} = \dfrac{3}{x}$

d) $x = 7$ e) $y = |x|$ f) $y = x^2 + 5$

2. Which are symmetric with respect to the origin?

    a) a, c, f   b) b, d   c) b, c   d) a, b, c

2. _____

3. Which are symmetric with respect to the x-axis?

    a) b, e, f   b) b, d   c) a, b, c   d) b, c

3. _____

4. Which are symmetric with respect to the y-axis?

    a) c,e   b) a, c, d   c) b, e, f   d) b, f

4. _____

5. Classify this function as even, odd, or neither even nor odd; $f(x) = 4x^7 - 3x^3$.

    a) Even   b) Odd   c) Neither

5. _____

6. Find the domain: $f(x) = \dfrac{3x + 1}{x^3 - 4x}$.

    a) $\{x \mid x \neq 0\}$ b) $\{x \mid x \neq -2,2\}$

    c) $\{x \mid x \neq 0, x \neq -2, x \neq 2\}$ d) $\{x \mid x \neq 2\}$

6. _____

NAME _____

7. _____

8. _____

9. _____

10. _____

11. _____

7. Given $f(x) = 3x - 2$ and $g(x) = x^2 + x - 1$, find $f \circ g(x)$.

a) $3x^2 + 3x - 5$

b) $9x^2 - 11x + 3$

c) $3x^2 + 3x - 3$

d) $9x^2 - 9x + 1$

8. Which of the following are graphs of functions?

   a)   b)   c)   d)

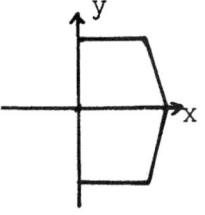 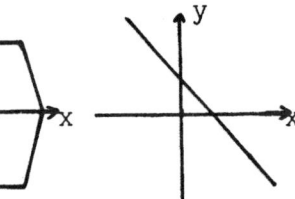 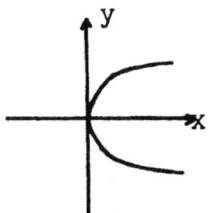 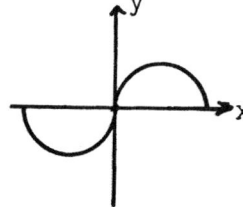

   a) b, d   b) a, d   c) b   d) b, c

9. Find $\dfrac{f(a + h) - f(a)}{h}$ when $f(x) = \dfrac{1}{x}$.

   a) $\dfrac{1}{a(a + h)}$ b) $\dfrac{1}{h^2}$ c) $\dfrac{-h^2}{a(a + h)}$ d) $\dfrac{-1}{a(a + h)}$

10. Find the slope of the line $3 - 4y = -6x$.

   a) -6 b) -4 c) $\dfrac{3}{2}$ d) $-\dfrac{2}{3}$

11. Find the distance between $(1,-4)$ and the midpoint of the segment with endpoints $(2,3)$ and $(-4,5)$.

   a) 4 b) $2\sqrt{10}$ c) $2\sqrt{13}$ d) $2\sqrt{17}$

				ANSWERS

12. Graph: $x = |y - 2|$.

ANSWERS

12. _____

a)

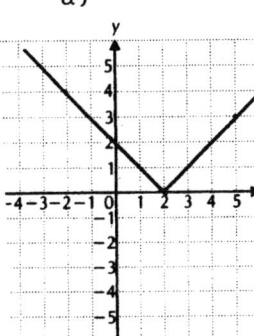

b)

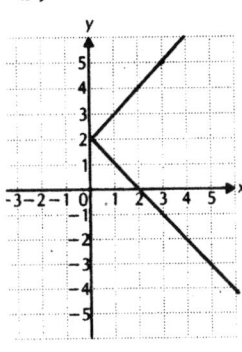

c)

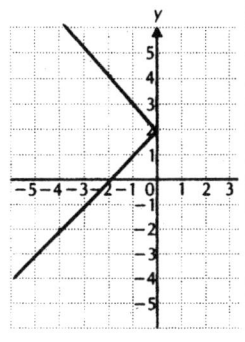

13. _____

13. Graph: $y = 1 - x^2$.

a)

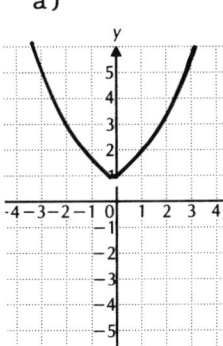

b)

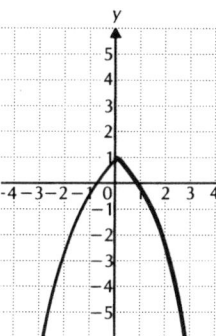

c)

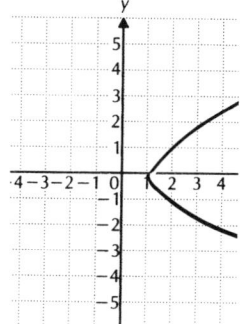

14. _____

14. Find an equation of the line containing the given point and parallel to the given line.

$(-2,1)$; $2y - 3x = 5$

a) $y = \frac{3}{2}x - \frac{3}{2}$　　　　　b) $y = \frac{2}{3}x - 2$

c) $y = \frac{3}{2}x + 4$　　　　　d) $y = \frac{3}{2}x - \frac{7}{2}$

15. _____

15. Write interval notation for $\left\{x \mid 0 \le x < \frac{5}{2}\right\}$.

a) $\left[0, \frac{5}{2}\right]$　　b) $\left(0, \frac{5}{2}\right)$　　c) $\left(0, \frac{5}{2}\right]$　　d) $\left[0, \frac{5}{2}\right)$

16. Classify this function as increasing, decreasing, or neither increasing nor decreasing.

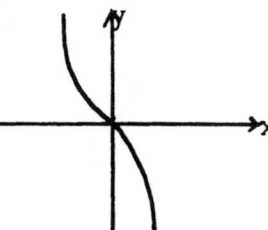

a) Increasing

b) Decreasing

16. _____

c) Neither

17. Here is a graph of $y = f(x)$.
Sketch the graph of $y = f(x) - 3$.

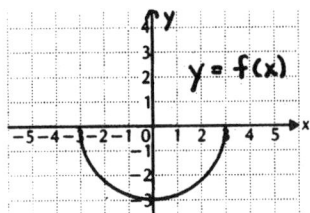

a)

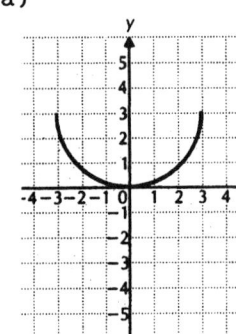

b)

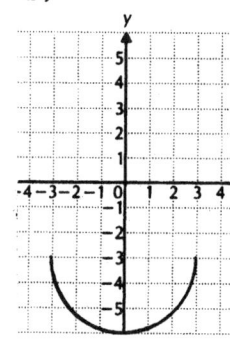

c)

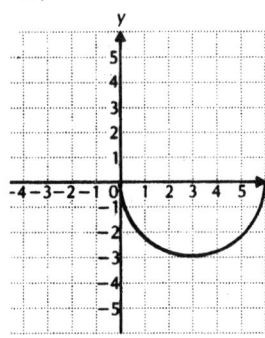

18. Here is a graph of $y = f(x)$.
Sketch the graph of $y = -\frac{1}{2}f(x + 1)$.

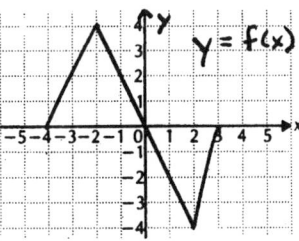

a)

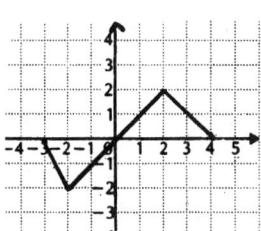

b)

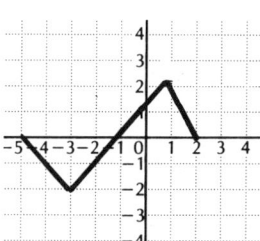

c)

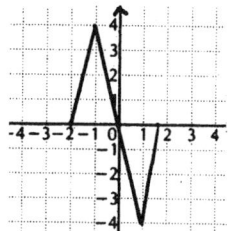

NAME _____

CLASS _____ SCORE _____ GRADE _____

For the functions in Questions 1 and 2:

a) use completing the square to put each equation into the form $f(x) = a(x - h)^2 + k$;

b) find the vertex; and

c) determine whether there is a maximum or minimum function value and find that value.

1. $f(x) = 2x^2 - 8x + 1$ 2. $f(x) = -3x^2 + 2x - 1$

3. Graph $f(x) = -2x^2 + 4x + 3$.

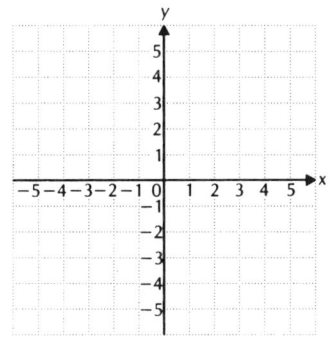

4. Find the x-intercepts of $f(x) = 4x^2 - 3x - 2$.

5. Find $\{2,7,9,11\} \cup \{1,2,5,9\}$.

6. Graph $\{x|x \leq 5\} \cap \{x|x > 3\}$.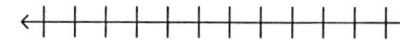

Solve.

7. $x + 2 > -6$ and $-3x + 1 \geq 4$ 8. $|2x - 5| \geq 7$

9. $|x - 7| < 3$ 10. $|3x - 2| = 8$

11. $x^2 - 10x + 9 > 0$ 12. $3x^2 + x - 2 < 0$

13. $\dfrac{x - 3}{x + 4} < 2$

ANSWERS

1. a) _____

   b) _____

   c) _____

2. a) _____

   b) _____

   c) _____

3. _____ See graph.

4. _____

5. _____

6. _____ See graph.

7. _____

8. _____

9. _____

10. _____

11. _____

12. _____

13. _____

NAME _____

ANSWERS

14. Find two numbers whose sum is -16 and whose product is a maximum.

14. _____

Sketch a graph of the polynomial function.

15. $f(x) = x^4 - 4x^2$

16. $f(x) = x^4 - 4x^2 + 2$

15. See graph.

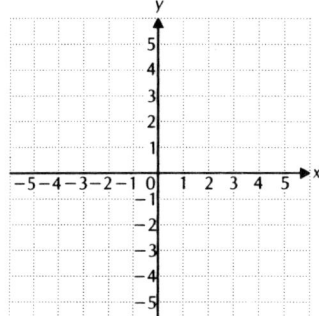

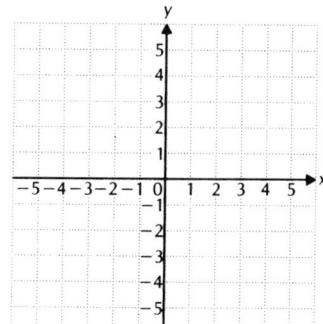

16. See graph.

17. _____

17. Find the remainder when $2x^4 - 3x^2 + 9x - 1$ is divided by $x - 4$.

18. _____

18. Determine whether $x - i$ is a factor of $2x^3 + 2x$.

19. _____

19. Use synthetic division to find the quotient and the remainder. Show all your work.

$$(4x^3 - 3x^2 + 2x - 2) \div (x - 1)$$

20. _____

20. Use synthetic division to find $P(-2)$.

$$P(x) = 3x^4 - 6x^3 + 2x^2 - 4$$

TEST FORM A

---

21. Factor the polynomial $P(x)$. Then solve the equation $P(x) = 0$.

$$P(x) = x^3 - 2x^2 - 11x + 12$$

22. Find a polynomial of degree 4 with roots 3, -3, $1 + i$, and $1 - i$.

23. Find a polynomial of lowest degree having roots 2 and -3, and having 1 as a root of multiplicity 2 and 5 as a root of multiplicity 3.

24. Find a polynomial of lowest degree with rational coefficients that has $4 - i$ and 3 as two of its roots.

25. List all possible rational roots of
$$4x^5 - 2x^3 - 3x^2 + 6.$$

26. What does Descartes' rule of signs tell you about the number of positive real roots and the number of negative real roots of the polynomial?
$$8x^6 + 3x^4 - 9x^3 + x - 7$$

ANSWERS
21. _____
22. _____
23. _____
24. _____
25. _____
26. _____

NAME _____

27. _____

28. _____

29. See graph. _____

30. _____

31. _____

32. _____

27. Find the smallest positive integer that is guaranteed by Theorem 15 to be an upper bound to the roots of
$$4x^5 - 3x^3 + 2x - 8.$$
Then find the largest negative integer that is guaranteed to be a lower bound.

28. Approximate the irrational roots of $P(x) = x^4 - 5x^2 - 1$.

29. Graph: $f(x) = \dfrac{3x + 1}{x}$.

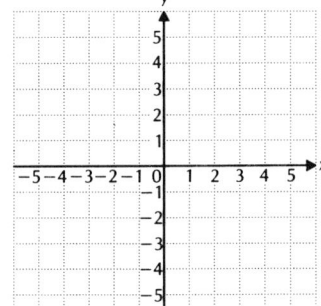

30. Find k such that $f(x) = kx^2 + 5x - 2$ has a minimum value at $x = -3$.

31. Solve: $|x + 4| \leq |x - 1|$.

32. Solve: $x^3 - 5x^2 - 2x + 24 > 0$.

For the functions in Questions 1 and 2:

a) use completing the square to put each equation into the
   form $f(x) = a(x - h)^2 + k$;

b) find the vertex; and

c) determine whether there is a maximum or minimum function
   value and find that value.

1. $f(x) = -3x^2 + 12x - 4$

2. $f(x) = 5x^2 + 2x + 1$

3. Graph $f(x) = 3x^2 + 6x + 5$.

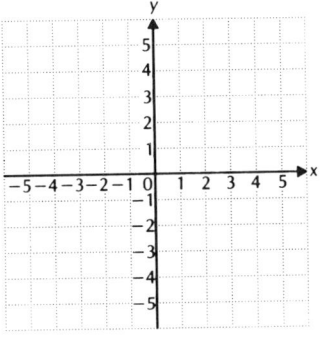

4. Find the x-intercepts of $f(x) = 2x^2 + 5x - 9$.

5. Find $\{2,5,7,11,15\} \cap \{5,10,15,20\}$.

6. Graph $\{x \mid x \leq 2\} \cup \{x \mid x > 4\}$.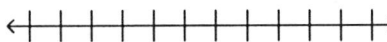

Solve.

7. $x - 1 > -5$ and $-2x - 5 \geq 7$

8. $|-3x + 1| > 4$

9. $|2x - 1| \leq 5$

10. $|5x - 1| = 21$

11. $x^2 + 2x - 8 > 0$

12. $10x^2 + 3x - 1 < 0$

13. $\dfrac{x - 1}{3x + 2} > 1$

ANSWERS

1. a) _____

   b) _____

   c) _____

2. a) _____

   b) _____

   c) _____

3. ___ See graph. ___

4. _____

5. _____

6. ___ See graph. ___

7. _____

8. _____

9. _____

10. _____

11. _____

12. _____

13. _____

NAME _____

ANSWERS

14. _____

15. _____ See graph.

16. _____ See graph.

17. _____

18. _____

19. _____

20. _____

14. Of all the numbers whose difference is 18, find the two that have the minimum product.

Sketch a graph of the polynomial function.

15. $f(x) = x^4 - 3x^2$

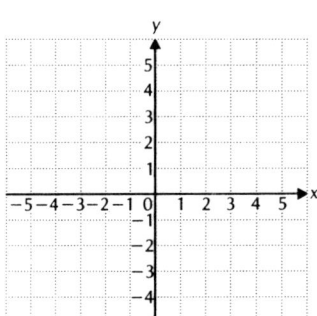

16. $f(x) = x^4 - 3x^2 + 4$

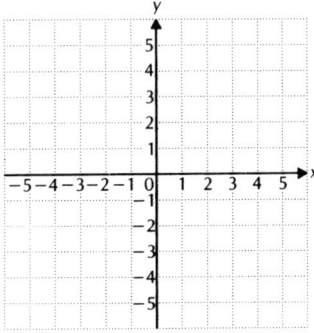

17. Find the remainder when $7x^5 - 2x^3 + x^2 - 2$ is divided by $x - 1$.

18. Determine whether $x - 2$ is a factor of $x^4 - 3x^2 - 4$.

19. Use synthetic division to find the quotient and the remainder. Show all your work.
$$(3x^4 - 2x^2 + 3x - 1) \div (x + 2)$$

20. Use synthetic division to find P(4):
$$P(x) = 2x^4 - 3x^3 + 2x - 5.$$

NAME _____

---

21. Factor the polynomial P(x). Then solve the equation P(x) = 0.

$$P(x) = x^3 - 2x^2 - 13x - 10$$

ANSWERS

21. _____

22. Find a polynomial of degree 3 with roots 5, $\sqrt{3}$, and $-\sqrt{3}$.

22. _____

23. Find a polynomial of lowest degree having roots 0 and 5, and having 3 as a root of multiplicity 2 and −1 as a root of multiplicity 4.

23. _____

24. Find a polynomial of lowest degree with rational coefficients that has 7 and $3 + \sqrt{2}$ as two of its roots.

24. _____

25. List all possible rational roots of $2x^6 - 5x^3 + 8x^2 - 8$.

25. _____

26. What does Descartes' rule of signs tell you about the number of positive real roots and the number of negative real roots of the polynomial?

$$-5x^6 + 3x^4 - x^3 + 2x + 5$$

26. _____

NAME _____

27. _____

28. _____

29. See graph.

30. _____

31. _____

32. _____

27. Find the smallest positive integer that is guaranteed by Theorem 15 to be an upper bound to the roots of

$$5x^4 - 3x^2 + 7x - 2.$$

Then find the largest negative integer that is guaranteed to be a lower bound.

28. Approximate the irrational roots of $P(x) = x^4 - x^2 - 7$.

29. Graph: $f(x) = \dfrac{x^2 - x - 1}{x - 2}$.

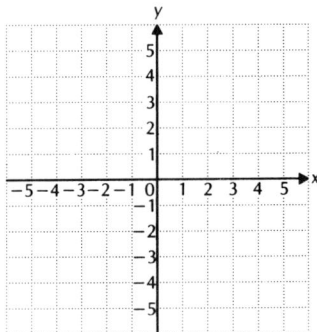

30. Find c such that $f(x) = -3x^2 + 6x + c$ has a maximum value of 39.

31. Solve: $x^4 - 5x^2 \le 0$.

32. When $x^2 - 5x + 2k$ is divided by $x + 3$, the remainder is -2. Find the value of k.

NAME _____

CLASS _____ SCORE _____ GRADE _____

For the functions in Questions 1 and 2:

a) use completing the square to put each equation into the form $f(x) = a(x - h)^2 + k$;

b) find the vertex; and

c) determine whether there is a maximum or minimum function value and find that value.

1. $f(x) = 6x^2 + 12x - 5$

2. $f(x) = -2x^2 + x + 3$

3. Graph $f(x) = -x^2 - 6x - 5$.

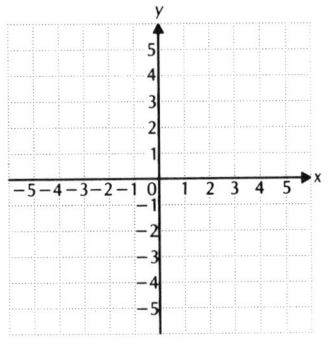

4. Find the x-intercepts of $f(x) = 3x^2 - 7x - 1$.

5. Find $\{1,4,7,8,16,20\} \cup \{1,7,9,11,18\}$.

6. Graph $\{x \mid x < 5\} \cap \{x \mid x \geq 0\}$. ←|||||||||||||||→

Solve.

7. $2x + 3 \geq 1$ and $5x - 1 < 4$

8. $|5x - 3| \geq 7$

9. $|x + 12| < 4$

10. $|8x - 3| = 27$

11. $x^2 + 4x - 12 > 0$

12. $12x^2 - 5x - 2 < 0$

13. $\dfrac{x - 5}{x + 1} > 3$

ANSWERS

1. a) _____

   b) _____

   c) _____

2. a) _____

   b) _____

   c) _____

3. _____See graph._____

4. _____

5. _____

6. _____See graph._____

7. _____

8. _____

9. _____

10. _____

11. _____

12. _____

13. _____

NAME _____

ANSWERS	
14. _____	14. The sum of the base and height of a triangle is 28 cm. Find the dimensions for which the area is a maximum.

Sketch a graph of the polynomial function.

15. $f(x) = x^4 - 2x^2$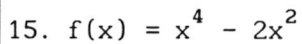

16. $f(x) = x^4 - 2x^2 - 3$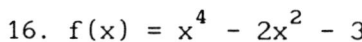

15. _See graph._

16. _See graph._

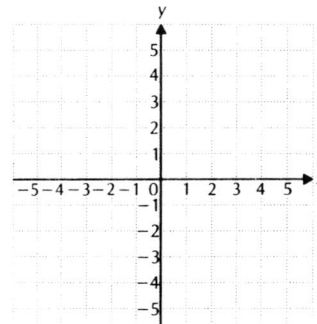   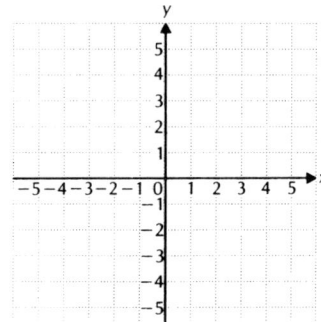

17. _____

17. Find the remainder when $2x^4 - 4x^3 - 2x^2 + 1$ is divided by $x + 2$.

18. _____

18. Determine whether $x - i$ is a factor of $x^4 + x$.

19. _____

19. Use synthetic division to find the quotient and the remainder. Show all your work.

$$(5x^3 - 7x^2 + 3x - 4) \div (x - 4)$$

20. _____

20. Use synthetic division to find $P(-3)$:

$$P(x) = 5x^4 - 2x^2 - 9x + 7.$$

---

21. Factor the polynomial P(x). Then solve the equation
    P(x) = 0.
    $$P(x) = x^3 + x^2 - 34x + 56$$

ANSWERS

21. _____

22. Find a polynomial of degree 4 with roots 4, −4, 3i, and
    −3i.

22. _____

23. Find a polynomial of lowest degree having roots −4 and
    −3, and having 5 as a root of multiplicity 2 and −2 as
    a root of multiplicity 3.

23. _____

24. Find a polynomial of lowest degree with rational

    coefficients that has −2 and −3 + 5i as two of its

    roots.

24. _____

25. List all possible rational roots of
    $6x^5 - 3x^3 + 2x^2 - 4$.

25. _____

26. What does Descartes' rule of signs tell you about the
    number of positive real roots and the number of
    negative real roots of the polynomial?
    $$6x^5 - 2x^4 + 3x^2 + 2x + 7$$

26. _____

NAME _____

ANSWERS	
27. _____	27. Find the smallest positive integer that is guaranteed by Theorem 15 to be an upper bound to the roots of $$3x^4 - 2x^3 + 7x - 1.$$ Then find the largest negative integer that is guaranteed to be a lower bound.
28. _____	28. Approximate the irrational roots of $$P(x) = x^4 - x - 3.$$
29. See graph.	29. Graph: $f(x) = \dfrac{x - 1}{x^2 - 3x - 10}$. 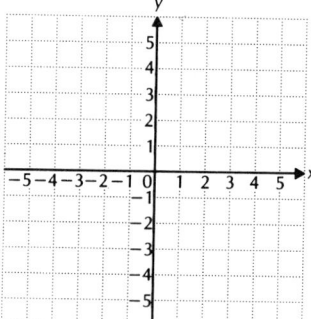
30. _____	30. Solve: $(x - 2)^2 > x(x - 5)$.
31. _____	31. The base of a triangle is 6 cm greater than the height. Find the possible heights h such that the area of the triangle will be greater than 15 cm.
32. _____	32. Find k so that $x - 3$ is a factor of $x^3 + kx^2 - 2x + 5k$.

NAME _____

CLASS _____ SCORE _____ GRADE _____

ANSWERS

For the functions in Questions 1 and 2:

a) use completing the square to put each equation into the form $f(x) = a(x - h)^2 + k$;

b) find the vertex; and

c) determine whether there is a maximum or minimum function value and find that value.

1. $f(x) = -2x^2 - 8x + 3$

2. $f(x) = 4x^2 + 5x - 2$

3. Graph $f(x) = 2x^2 - 4x + 5$.

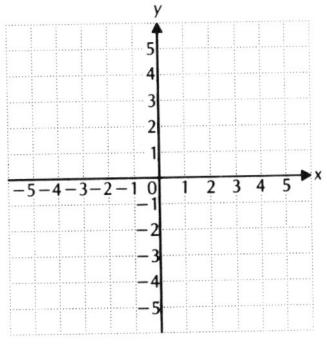

1. a) _____

   b) _____

   c) _____

2. a) _____

   b) _____

   c) _____

3. ___See graph.___

4. _____

5. _____

6. ___See graph.___

7. _____

8. _____

4. Find the x-intercepts of $f(x) = 3x^2 + 5x - 7$.

5. Find $\{1,4,7,8,16,20\} \cap \{1,7,9,11,18\}$.

6. Graph $\{x|x < -3\} \cup \{x|x > -1\}$. ⟵+++++++++++++++⟶

9. _____

Solve.

7. $-3x + 1 < 7$ and $2x + 3 < 9$ 8. $|7 - x| > 9$

9. $|3x + 2| \le 5$ 10. $|4x + 5| = 11$

11. $x^2 - 2x - 24 > 0$ 12. $7x^2 - 27x - 4 < 0$

13. $\dfrac{2x - 1}{x + 7} \le 1$

10. _____

11. _____

12. _____

13. _____

NAME _____

14. _____

14. Of all the numbers whose difference is 22, find the two that have the minimum product.

Sketch a graph of the polynomial function.

15. $f(x) = x^4 - 5x^2$

16. $f(x) = x^4 - 5x^2 + 4$

15. _____See graph._____

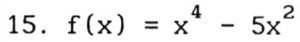

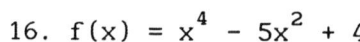

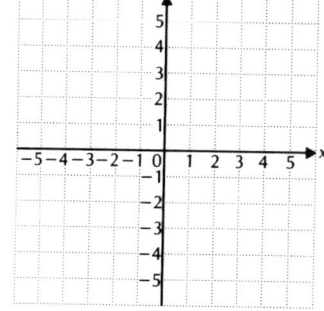

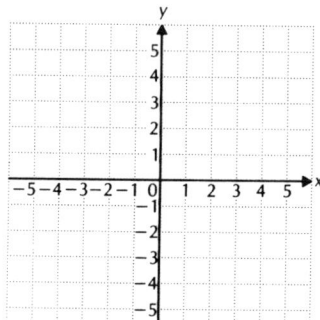

16. _____See graph._____

17. _____

17. Find the remainder when $3x^4 - 2x^2 + x + 3$ is divided by $x + 3$.

18. _____

18. Determine whether $x - 3$ is a factor of $x^3 - 9$.

19. _____

19. Use synthetic division to find the quotient and the remainder. Show all your work.

$$(x^4 + 2x^3 - x + 4) \div (x + 3).$$

20. _____

20. Use synthetic division to find P(5):

$$P(x) = 4x^4 - 2x^3 + x - 9.$$

NAME _____

	ANSWERS
21. Factor the polynomial P(x). Then solve the equation P(x) = 0. $$P(x) = x^3 - 7x^2 + 7x + 15$$	21. _____
22. Find a polynomial of degree 3 with roots 7, $2 + \sqrt{5}$, and $2 - \sqrt{5}$.	22. _____
23. Find a polynomial of lowest degree having roots 1 and −2, and having −1 as a root of multiplicity 2 and 6 as a root of multiplicity 3.	23. _____
24. Find a polynomial of lowest degree with rational coefficients that has $1 - \sqrt{5}$ and 4 as two of its roots.	24. _____
25. List all possible rational roots of $8x^5 - 3x^3 + 2x - 4.$	25. _____
26. What does Descartes' rule of signs tell you about the number of positive real roots and the number of negative real roots of the polynomial? $$10x^6 - 5x^5 + 3x^4 - 2x^3 - x + 5$$	26. _____

ANSWERS	

27. _____

27. Find the smallest positive integer that is guaranteed by Theorem 15 to be an upper bound to the roots of
$$6x^5 - 3x^2 + 2x - 15.$$
Then find the largest negative integer that is guaranteed to be a lower bound.

28. _____

28. Approximate the irrational roots of
$$P(x) = 2x^3 + x^2 - 2.$$

29. Graph: $f(x) = \dfrac{x^2 - 2x - 3}{x + 2}$.

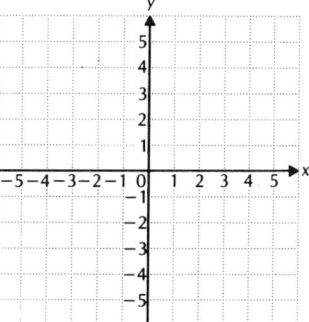

29. See graph.

30. _____

30. Find the domain of the function: $f(x) = \dfrac{\sqrt{x + 5}}{\sqrt{x - 5}}$.

31. A company has the following total-cost and total-revenue functions to use in producing and selling x units of a certain product.
$$R(x) = 36x - x^2, \quad C(x) = 4x + 220$$

31. a) _____

    b) _____

a) Find the break-even values $(R(x) = C(x))$.

b) Find the values of x that produce a profit.

32. When $x^2 - 5x + 2k$ is divided by x + 3, the remainder is 10. Find the value of k.

32. _____

NAME _____

CLASS _____ SCORE _____ GRADE _____

1. For the function $f(x) = 2x^2 - 10x - 5$,

a) find an equation of the type $f(x) = a(x - h)^2 + k$,

b) find the vertex; and

c) determine whether there is a maximum or minimum function value and find that value.

2. Graph: $f(x) = -x^2 + 3x - 1$.
   Also find the x-intercepts.

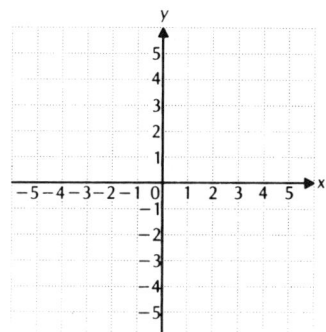

3. The sum of the length and width of a rectangle is 54. Find the dimensions for which the area is a maximum.

4. Find $\{b,d,f,h,j\} \cup \{d,h,\ell,p\}$.

5. Graph: $\{x \mid x < -\frac{3}{2}\} \cup \{x \mid x \geq -1\}$.

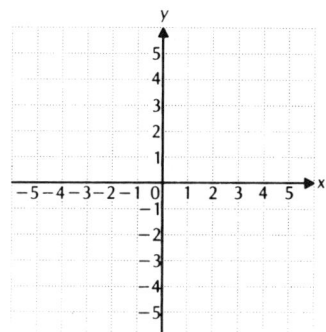

ANSWERS

1. a) _____

   b) _____

   c) _____

2. _____See graph._____

3. _____

4. _____

5. _____See graph._____

ANSWERS	Matching: Solve and place the letter of the answer in the answer blank. Some letters may be used more than once, and some may not be used.

6. _____

7. _____

8. _____

9. _____

10. _____

11. _____

12. _____

13. See graph.

Solutions

6. $|8 - 2x| = 4$ a) $(3,7)$ b) $\left[\frac{5}{6}, 2\right]$

7. $|-3x + 7| \geq 19$ c) $(-\infty, -1) \cup (7, \infty)$

8. $x - 1 > -7$ and   d) $\{-2, 4\}$ e) $(-6, -2]$
   $-7x + 2 \geq 16$

        f) $(-1, 7)$ g) $\left(-4, -\frac{3}{2}\right)$

9. $x^2 - 6x - 7 < 0$

        h) $(-\infty, -4] \cup \left[\frac{26}{3}, \infty\right)$

10. $|x - 5| < 2$

11. $6x^2 - 17x + 10 \leq 0$ i) $\{-6, -\frac{4}{3}\}$ j) $\{2, 6\}$

12. $\dfrac{x - 1}{2x + 3} > 1$ k) $\left[4, \frac{26}{3}\right]$ ℓ) $\varnothing$

Sketch a graph of each polynomial function.

13. $f(x) = x^3 - 3x^2 + x + 1$

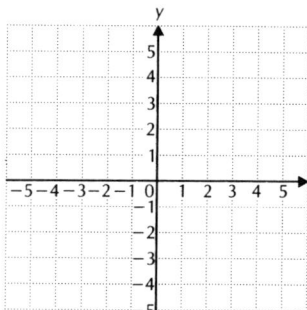

NAME 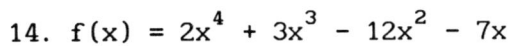 _____

	ANSWERS

Sketch a graph of each polynomial function.

14. $f(x) = 2x^4 + 3x^3 - 12x^2 - 7x + 6$

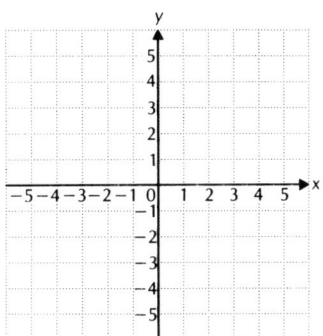

14. __See graph.__

15. Find the remainder when $P(x) = x^2 - 5ix + 4$ is divided by $x - i$.

15. _____

16. Use synthetic division to find the quotient and remainder. Show all your work.

$$(5x^5 - 2x^3 + x^2 - 2x + 3) \div (x + 2)$$

16. _____

17. _____

17. Use synthetic division to find $P(-5)$.

$$P(x) = x^4 - 2x^3 + 5x - 16$$

18. _____

18. Factor the polynomial $P(x)$. Then solve the equation $P(x) = 0$.

$$P(x) = x^5 + 3x^4 - 26x^3 - 78x^2 + 25x + 75$$

19. _____

19. Find a polynomial of degree 4 with roots $1 - \sqrt{3}$, $1 + \sqrt{3}$, $2i$, and $-2i$.

20. Find a polynomial of lowest degree having roots 0 and -2, and having 5 as a root of multiplicity 4 and -7 as a root of multiplicity 3.

20. _____

NAME _____

ANSWERS
21. _____

21. Given that $x^3 - 64 = 0$ has 4 as a root, find the other roots.

22. A polynomial of degree 7 with rational coefficients has 4, $-2i$, $\sqrt{7}$, and $4 - 5i$ as roots. Find the other roots.

22. _____

23. List all possible rational roots of $6x^3 - 5x + 4$.

24. What does Descartes' rule of signs tell you about the number of negative real roots of the polynomial?
$$-x^4 - 3x^3 + x^2 + 5x + 1$$

23. _____

25. Find the smallest positive integer that is guaranteed by Theorem 15 to be an upper bound to the roots of
$$2x^5 - 3x^3 + x - 5.$$

Then find the largest negative integer that is guaranteed to be a lower bound.

24. _____

26. Graph: $y = \dfrac{x^2 + 2x - 8}{x - 1}$.

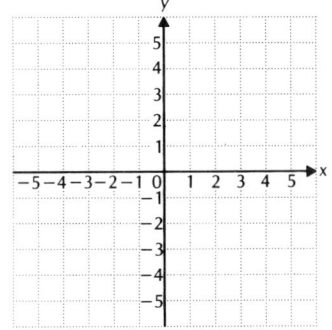

25. _____

26. See graph.

NAME _____

CLASS _____ SCORE _____ GRADE _____

ANSWERS

1. Determine whether there is a maximum or minimum function value of the function $f(x) = 4x^2 - 8x - 3$ and find that value.

   a) 7 is a minimum   b) -2 is a maximum

   c) -2 is a maximum   d) -7 is a minimum

1. _____

2. Find the x-intercepts of $f(x) = x^2 - 5x + 4$.

   a) $(1 + \sqrt{2}, 0)$, $(0, 1 - \sqrt{2})$

   b) $(1, 0)$, $(4, 0)$

   c) $(5, 0)$, $(-1, 0)$

   d) No x-intercepts

2. _____

3. Find $\{3, 6, 9, 12, 15, 18\} \cap \{5, 10, 15, 20, 25\}$.

   a) $\{15\}$ b) $\{3, 6, 9, 10, 12, 15, 20, 25\}$

   c) $\{3, 6, 9, 12, 18\}$ d) $\{15, 20, 25\}$

3. _____

4. Graph: $\{x | 2x - 3 \geq 1\} \cup \{x | -x + 1 > -2\}$.

   a)

   b)

   c)

   d)

4. _____

5. Solve: $|-4x + 3| > 15$.

   a) $\left(\frac{9}{2}, 13\right)$ b) $(-\infty, -3) \cup \left(\frac{9}{2}, \infty\right)$

   c) $\left(-3, \frac{9}{2}\right)$ d) $(-3, -2)$

5. _____

6. Solve: $|x - 5| \leq 2$.

   a) $[3, 7]$ b) $(-\infty, 3] \cup [7, \infty)$

   c) $[-3, 7]$ d) $[2, 5]$

6. _____

NAME _____

7. _____

7. Solve: $|6x + 3| = 15$.

    a) $(-\infty, -3] \cup [2, \infty)$ b) $(-3, 2]$

    c) $\{-3, 2\}$ d) $\{-5, 1\}$

8. _____

8. Solve: $2x - 5 < -2x + 3$.

    a) $(-2, \infty)$ b) $(-2, 2)$ c) $(-\infty, 4)$ d) $(-\infty, 2)$

9. _____

9. Solve: $x^2 - 5x - 14 > 0$.

    a) $(-3, 9)$ b) $(-7, 2)$

    c) $(-\infty, -2) \cup (7, \infty)$ d) $(-2, 7)$

10. _____

10. Solve: $3x^2 - 10x - 8 \leq 0$.

    a) $\left(-\dfrac{3}{2}, 4\right]$ b) $\left[-\dfrac{2}{3}, 4\right]$

    c) $\left(-\infty, -\dfrac{2}{3}\right] \cup (4, \infty)$ d) $(-\infty, -4) \cup \left[\dfrac{3}{2}, \infty\right)$

11. _____

11. Solve: $\dfrac{x + 3}{x} > 2$.

    a) $(0, 3)$ b) $(-\infty, 0) \cup (3, \infty)$

    c) $(2, 3)$ d) $(-\infty, 1) \cup (3, \infty)$

12. _____

12. Find one of two numbers whose sum is -12 and whose product is a maximum.

    a) -7 b) -10 c) -4 d) -6

	ANSWERS
13. Find the remainder when $2x^5 - x^3 + 4x^2 - x + 4$ is divided by $x + 2$. a) -42 b) 44 c) -34 d) 74	13. _____
14. Use synthetic division to find the quotient and remainder. $$(x^4 + 3x^2 - 2x + 1) \div (x - 3)$$ a) Q: $x^2 + 9x + 25$, R: 76 b) Q: $x^3 + 3x^2 + 12x + 34$, R: 103 c) Q: $x^3 + 6x^2 + 18x + 52$, R: 157 d) Q: $x^3 - 3x^2 + 12x - 38$, R: 115	14. _____
15. Factor the polynomial P(x). Then solve the equation P(x) = 0. $$P(x) = x^5 - x^4 - 8x^3 + 8x^2 - 9x + 9$$ Find the positive real roots. a) 1 and 2 b) 1 and 4 c) 1 and 3 d) 3 only	15. _____
16. Find a polynomial of degree 4 with roots $\sqrt{5}$, $-\sqrt{5}$, $2 - i$, and $2 + i$. a) $x^4 - 4x^3 + 20x - 25$ b) $x^4 + 2x^3 - 4x^2 + 15x + 5$ c) $x^4 - 25$ d) $x^4 - 4x^3 - 10x^2 + 20x - 25$	16. _____

17. _____

18. _____

19. _____

20. _____

17. Find a polynomial of lowest degree having roots -3 and 1, and having 0 as a root of multiplicity 3 and -4 as a root of multiplicity 2.

   a) $3x(x + 3)(x - 1)(x - 4)^2$

   b) $x^3(x - 3)(x + 1)(x - 4)^2$

   c) $(x + 3)(x - 1)(x + 4)$

   d) $x^3(x + 3)(x - 1)(x + 4)^2$

18. A polynomial of degree 5 with rational coefficients has 4, $2 + \sqrt{7}$, and $-3 - 2i$ as roots. Find the other roots.

   a) $-4, -2 - \sqrt{7}, 3 + 2i$　　　b) $2 - \sqrt{7}, -3 + 2i$

   c) $-2 - \sqrt{7}, 3 + 2i$　　　　　d) $-2 + \sqrt{7}, 3 - 2i$

19. Given that $x^3 - 8$ has 2 as a root, find the other roots.

   a) ± 2　　　b) $\pm 2i$　　　c) $-1 \pm \sqrt{3}i$　　d) $-1 \pm \sqrt{3}$

20. List all possible rational roots of $4x^2 - 5x - 8$.

   a) $\pm \left(\frac{1}{4}, \frac{1}{2}, 1, 2, 4, 8\right)$　　　b) $\pm (1, 2, 4, 8)$

   c) $\pm \left(\frac{1}{4}, \frac{1}{2}, \frac{1}{8}, 1, 2, 4\right)$　　　d) $\pm (1, 2, 4)$

NAME _____

CLASS _____ SCORE _____ GRADE _____

ANSWERS

1. Find the inverse of the relation G given by

    G = {(2,−1), (5,−3), (7,1), (−4,2), (3.2,4.5)}.

2. Write an equation of the inverse of the relation

    y = 3x − 4.

3. Which of the following have inverses that are functions?

    a) 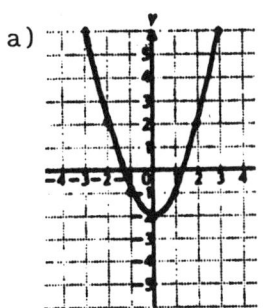   b) 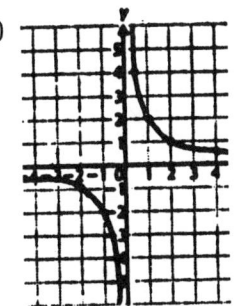   c)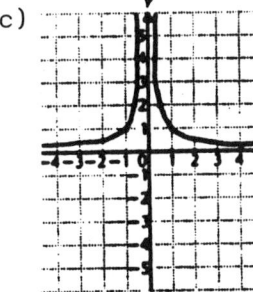

4. Find a formula for $f^{-1}(x)$:   f(x) = 6 − 2x.

5. Find $h(h^{-1}(3))$: $h(x) = 35x^2 - 19x$.

6. Graph: $y = \log_8 x$.   7. Graph: $f(x) = 3e^x$.

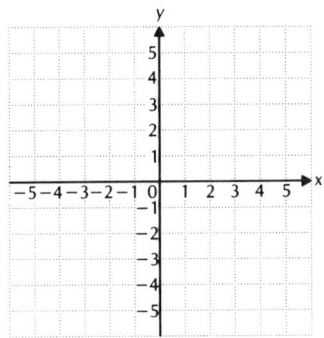   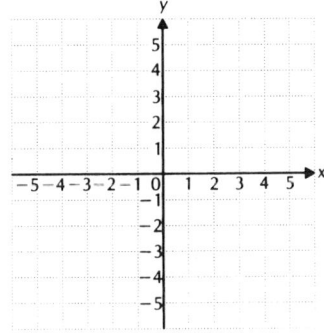

8. Find $\log_4 32$ using natural logarithms.

9. Write an exponential equation equivalent to $\log_{\sqrt{3}} 27 = 6$.

1. _____

2. _____

3. _____

4. _____

5. _____

6. _See graph._

7. _See graph._

8. _____

9. _____

NAME _____

ANSWERS	
10. _____	10. Write a logarithmic equation equivalent to $$x^4 = 0.2401.$$
11. _____	11. Simplify: $6^{\log_6 (x+2)}$.
12. _____	12. Write an equivalent expression containing a single logarithm: $$\frac{3}{4} \log_a x - \frac{1}{2} \log_a y + 5 \log_a z.$$
13. _____	13. Express in terms of logarithms of a and b: $$\log \frac{a^2}{b}.$$
14. _____	Given that $\log_a 2 = 0.301$, $\log_a 7 = 0.845$, and $\log_a 6 = 0.778$, find each of the following.
15. _____	
16. _____	14. $\log_a 3$ 15. $\log_a 28$ 16. $\log_a \sqrt[4]{6}$
17. _____	17. Solve for x: $\log_d d^{3x^2} = 6$.
18. _____	Solve.
19. _____	18. $\log_4 x = 3$
20. _____	19. $6^{2x+1} + 7 = 43$
21. _____	20. $\log_5 (x + 4) + \log_5 x = 1$
	21. $e^{-x} = 0.4$

NAME _____

Find using a calculator.	ANSWERS

22. log (-1.04)   23. ln 0.000013

22. _____

24. ln 2.05   25. log 0.04372

23. _____

26. log 49,500

24. _____

27. How many years will it take an investment of $3500 to double if interest is compounded annually at 10%?

25. _____

26. _____

28. An earthquake was measured at an intensity of $10^{6.34} \cdot I_0$. What was its magnitude on the Richter scale?

27. _____

29. The population of a city was 60,000 in 1980 and 80,000 in 1990. Estimate the population in 2000.

28. _____

30. The average walking speed R of people living in a city of population P, in thousands, is given by $R = 0.37 \ln P + 0.05$, where R is in feet per second.

    a) The population of Boston, Massachusetts, is 572,000. Find the average walking speed.

29. _____

    b) A city's population has an average walking speed of 3.1 ft/sec. Find the population.

30. a) _____

    b) _____

NAME _____

ANSWERS	
31. a) _____	31. The cost of a Sweet ice cream bar in 1970 was 15 cents and was increasing at an exponential growth rate of 8.4%.
b) _____	a) Find an exponential function describing the growth of the cost of a Sweet bar.
c) _____	b) What will a Sweet bar cost in 1995?
d) _____	c) When will a Sweet bar cost $3?
	d) What is the doubling time of the cost of a Sweet bar?
32. _____	32. The population of a city doubled in 20 years. What was the exponential growth rate?
33. _____	33. How long will it take an investment to double itself if it is invested at 6.8%, compounded continuously?
34. _____	34. How old is an animal bone that has lost 42% of its carbon-14?
35. _____	35. What is the loudness, in decibels, of a sound whose intensity is $550 \cdot I_0$?
36. _____	36. The hydrogen ion concentration of a substance is 2.4×10^{-5}. What is the pH?
37. _____	37. Determine whether $\pi^{4.2}$ or 4.2^{π} is larger?
38. _____	38. Solve: $5^{\log_5 (3x-7)} = 4$.

1. Find the inverse of the relation H given by

   H = {(8,-2), (6,-7), (8,3), (-5,1), (5.6,-1.2)}.

2. Write an equation of the inverse of the relation
   $y = x^2 + 1$.

3. Which of the following have inverses that are functions?

a) 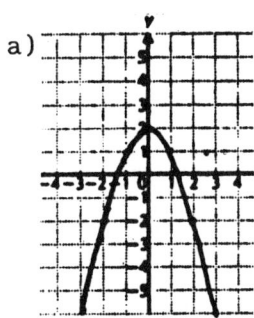 b) 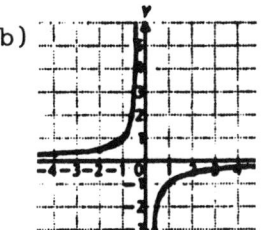 c)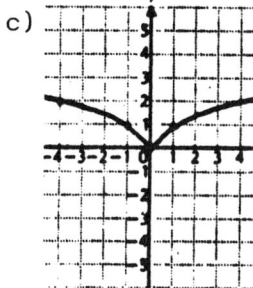

4. Find a formula for $f^{-1}(x)$: $f(x) = \dfrac{3}{x + 2}$.

5. Find $h(h^{-1}(-2))$: $h(x) = \dfrac{5x^2 - 1}{3}$.

6. Graph: $y = \log_2 x$.

7. Graph: $f(x) = e^{x+2}$.

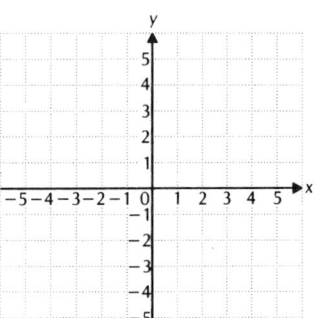

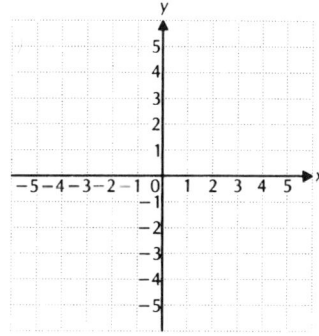

8. Find $\log_5 17$ using natural logarithms.

9. Write an exponential equation equivalent to
   $\log_{\sqrt{2}} 16 = 8$.

ANSWERS

1. _____

2. _____

3. _____

4. _____

5. _____

6. See graph. ___

7. See graph. ___

8. _____

9. _____

ANSWERS	
10. _____	10. Write a logarithmic equation equivalent to $$x^5 = 0.00243.$$
11. _____	11. Simplify: $5^{\log_5 4x}$.
12. _____	12. Write an equivalent expression containing a single logarithm: $$\frac{1}{2} \log_a x - 3 \log_a y + 4 \log_a z.$$
13. _____	13. Express in terms of logarithms of x and y: $$\log x^2 \sqrt{y}.$$
14. _____	
15. _____	Given that $\log_a 2 = 0.301$, $\log_a 5 = 0.699$, and $\log_a 8 = 0.903$, find each of the following.
16. _____	14. $\log_a 4$ 15. $\log_a 200$ 16. $\log_a \sqrt{2}$
17. _____	17. Solve for x: $\log_c c^{x^2+1} = 10$.
18. _____	
19. _____	Solve.

18. $\log_6 x = 2$ |
20. _____	19. $5^{3x-1} - 2 = 23$
21. _____	20. $\log_2 x + \log_2 (x + 6) = 4$
	21. $e^x = 1.2$

TEST FORM B

Find using a calculator.	ANSWERS

Find using a calculator.

22. log 25,400

23. ln 4.92

24. ln 0.000051

25. log 0.00324

26. log (-13.62)

27. How many years will it take an investment of $5000 to double if interest is compounded annually at 6%?

28. What is the loudness, in decibels, of a sound whose intensity is $6,000 \cdot I_0$?

29. The population of a city was 50,000 in 1970 and 75,000 in 1985. Estimate the population in 2000.

30. A model for advertising response is given by $N(a) = 3000 + 200 \log a$, $a \geq 1$, where $N(a)$ = the number of units sold and a = the amount spent on advertising, in thousands of dollars.

   a) How many units were sold after spending $1000 (a = 1) on advertising?

   b) How many units were sold after spending $8000?

   c) How much would have to be spent in order to sell 4000 units?

ANSWERS

22. _____

23. _____

24. _____

25. _____

26. _____

27. _____

28. _____

29. _____

30. a) _____

   b) _____

   c) _____

NAME _____

ANSWERS	
31. a) _____	31. The population of Bosnia was 4.5 million in 1984. The exponential growth rate was 3.2% per year.
b) _____	a) Find the exponential growth function.
c) _____	b) Predict the population of Bosnia in the year 2000.
	c) When will the population be 6.0 million?
32. _____	32. The population of a city doubled in 34.2 years. What was the exponential growth rate?
33. _____	33. How long will it take an investment to double itself if it is invested at 9.4%, compounded continuously?
34. _____	34. How old is an animal bone that has lost 58% of its carbon-14?
35. _____	35. An earthquake was measured at an intensity of $10^{2.54} \cdot I_0$. What was its magnitude on the Richter scale?
36. _____	36. The pH of a substance is 6.1. What is the hydrogen ion concentration?
37. a) _____	37. Approximate to six decimal places.
b) _____	a) 5^3 c) $5^{3.141}$
c) _____	b) $5^{3.14}$ d) $5^{3.1415}$
d) _____	
38. _____	38. Simplify: $\dfrac{\log_8 16}{\log_8 2}$.

NAME _____

CLASS _____ SCORE _____ GRADE _____

	ANSWERS

1. Find the inverse of the relation J given by

   J = {(-1,3), (4,-7), (6,0), (2,5), (-1.3,7.6)}.

2. Write an equation of the inverse of the relation

   $y = |x + 3|$.

3. Which of the following have inverses that are
   functions?

a)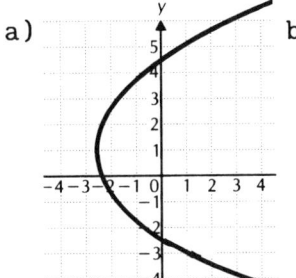
b)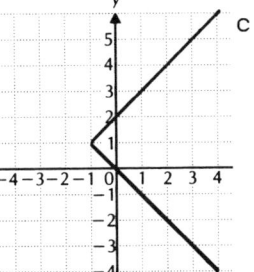
c)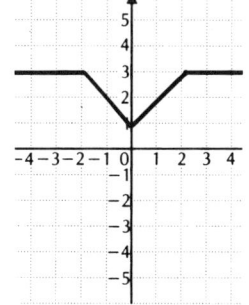

4. Find a formula for $f^{-1}(x)$: $f(x) = \dfrac{x + 2}{x - 1}$.

5. Find $h(h^{-1}(-2))$: $h(x) = \dfrac{25x - 7}{15}$.

6. Graph: $y = \log_5 x$.

7. Graph: $f(x) = 1.5e^x$.

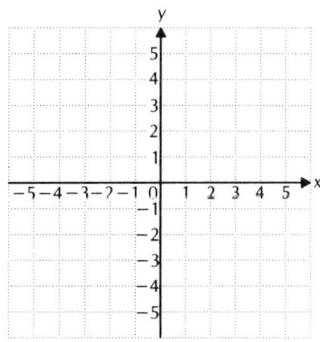

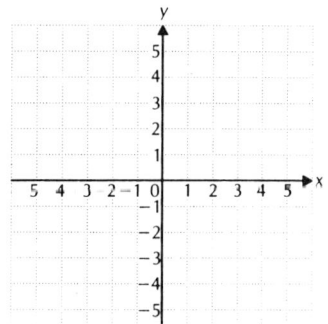

8. Find $\log_3 41$ using natural logarithms.

9. Write an exponential equation equivalent to
   $\log_{\sqrt{5}} 25 = 4$.

ANSWERS

1. _____

2. _____

3. _____

4. _____

5. _____

6. See graph. _____

7. See graph. _____

8. _____

9. _____

NAME _____

ANSWERS	
10. _____	10. Write a logarithmic equation equivalent to $x^6 = 0.015625$.
11. _____	11. Simplify: $8^{\log_8 (x-1)}$.
12. _____	12. Write an equivalent expression containing a single logarithm: $$5 \log_a x + \frac{1}{3} \log_a y - 2 \log_a z.$$
13. _____	13. Express in terms of logarithms of c and d: $$\log 5cd^2.$$
14. _____	
15. _____	Given that $\log_a 3 = 0.477$, $\log_a 4 = 0.602$, and $\log_a 6 = 0.778$, find each of the following.
16. _____	14. $\log_a 2$ 15. $\log_a 48$ 16. $\log_a \sqrt[3]{6}$
17. _____	17. Solve for x: $\log_b b^{x^2} = 4$.
18. _____	Solve.
19. _____	18. $\log_5 x = 3$
20. _____	19. $4^{5x-1} - 1 = 63$
21. _____	20. $\log_3 x + \log_3 (x - 8) = 2$
	21. $e^x = 0.9$

NAME _____

Find using a calculator.   **ANSWERS**

22. ln 0.0000021   23. log 101,200   22. _____

24. log 0.13724   25. log (-2.73)   23. _____

26. ln 5.12   24. _____

27. How many years will it take an investment of $2000 to   25. _____
    double if interest is compounded annually at 8%?

                                                     26. _____

28. An earthquake was measured at an intensity of
    $10^{5.17} \cdot I_0$. What was its magnitude on the Richter
    scale?

                                                     27. _____

29. The population of a city was 100,000 in 1960 and
    130,000 in 1970. Estimate the population in 2000.
                                                     28. _____

30. It is known that $\frac{1}{4}$ of aluminum cans distributed will
    be recycled each year. A beverage company distributes   29. _____
    300,000 cans. The number still in use after time t,
    in years, is given by the function
    $$N(t) = 300,000 \left(\frac{1}{4}\right)^t.$$

    a) After what amount of time will 50,000 still be in   30. a) _____
       use?
                                                         b) _____
    b) After what amount of time will only 100 cans still
       be in use?

NAME _____

ANSWERS

31. a) _____

b) _____

c) _____

31. The cost of a first-class postage stamp became 3 cents in 1932, and the exponential growth rate of the cost was 3.8% per year.

   a) Find the exponential growth function.

   b) Predict the cost of a first-class postage stamp in the year 2010.

   c) When will the cost of the first-class postage stamp be $1.50?

32. _____

32. The population of a city doubled in 18.4 years. What was the exponential growth rate?

33. _____

33. How long will it take an investment to double itself if it is invested at 10%, compounded continuously?

34. _____

34. How old is an animal bone that has lost 35% of its carbon-14?

35. _____

35. What is the loudness, in decibels, of a sound whose intensity is $4{,}350 \cdot I_0$?

36. _____

36. The pH of a substance is 4.9. What is the hydrogen ion concentration?

37. Solve: $\log_{64} x = \frac{2}{3}$.

37. _____

38. Solve: $\log_4 |x| = 2$.

38. _____

NAME _____

CLASS _____ SCORE _____ GRADE _____

ANSWERS

1. Find the inverse of the relation K given by

K = {(-4,3), (2,-5), (1,4), (6.2,4.3), (0,7)}.

1. _____

2. Write an equation of the inverse of the relation

xy = 3.

3. Which of the following have inverses that are functions?

2. _____

a) 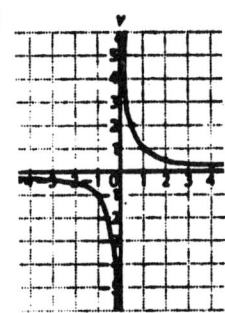 b) 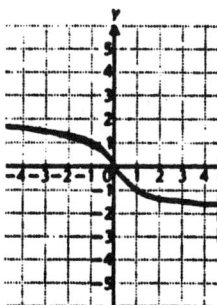 c)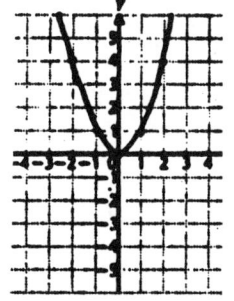

3. _____

4. Find a formula for $f^{-1}(x)$: $f(x) = \sqrt{x + 3}$.

4. _____

5. Find $h(h^{-1}(-7))$: $h(x) = \dfrac{x^2 - 7}{2x + 1}$.

5. _____

6. Graph: $y = \log_9 x$.   7. Graph: $f(x) = e^{2x-1}$.

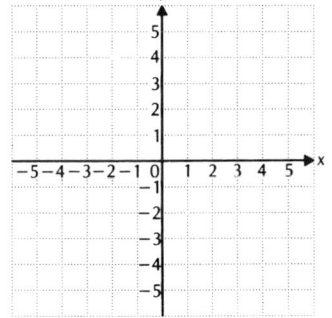   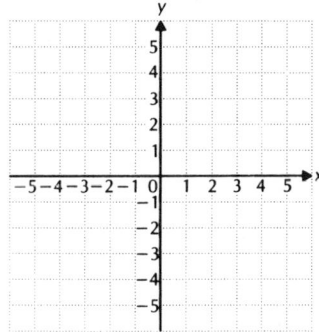

6. See graph.

7. See graph.

8. Find $\log_2 25$ using natural logarithms.

8. _____

9. Write an exponential equation equivalent to $\log_{\sqrt{5}} 125 = 6$.

9. _____

NAME _____

TEST FORM D

ANSWERS	

10. Write a logarithmic equation equivalent to
$$x^5 = 0.00032.$$

10. _____

11. Simplify: $7^{\log_7 5x}$.

11. _____

12. Write an equivalent expression containing a single logarithm:
$$2 \log_a x + 4 \log_a y - \frac{2}{3} \log_a z.$$

12. _____

13. Express in terms of logarithms of a and b:
$$\log \sqrt[3]{a^2 b}.$$

13. _____

14. _____

Given that $\log_a 4 = 0.602$, $\log_a 5 = 0.699$, and $\log_a 12 = 1.079$, find each of the following.

15. _____

14. $\log_a 3$ 15. $\log_a 100$ 16. $\log_a \sqrt{12}$

16. _____

17. Solve for x: $\log_a a^{3x^2} = 3$.

17. _____

Solve.

18. _____

18. $\log_2 x = 3$

19. _____

19. $2^{3x-5} - 3 = 13$

20. _____

20. $\log_4 x + \log_4 (x - 6) = 2$

21. _____

21. $e^{-x} = 0.7$

NAME _____

		ANSWERS
Find using a calculator.		

22. log 2.894   23. ln 6.23

24. log 17,350   25. log (-0.73)

26. ln 0.0000921

27. How many years will it take an investment of $8000 to double if interest is compounded annually at 11%?

28. What is the loudness, in decibels, of a sound whose intensity is $1,950I_0$?

29. The population of a city was 35,000 in 1960 and 42,000 in 1965. Estimate the population in 1990.

30. Students in a Spanish class took a final exam. They took equivalent forms of the exam in monthly intervals thereafter. The average score S(t), in percent, after t months was found to be given by

$$S(t) = 75 - 14 \log (t + 1), \quad t \geq 0.$$

a) What was the average score when they initially took the test, t = 0?

b) What was the average score after 6 months?

c) After what time t would the average score be 50?

ANSWERS

22. _____

23. _____

24. _____

25. _____

26. _____

27. _____

28. _____

29. _____

30. a) _____

b) _____

c) _____

NAME _____

ANSWERS	
31. a) _____	**31.** Under ideal conditions, a population increase of rats has an exponential growth rate of 18.2% per day. Suppose one starts with a population of 100 rats.
b) _____	a) Find the exponential growth function.
c) _____	b) What will the population of rats be after 10 days?
	c) What is the doubling time of the population of rats?
32. _____	**32.** The population of a city doubled in 36 years. What was the exponential growth rate?
33. _____	**33.** How long will it take an investment to double itself if it is invested at 8.7%, compounded continuously?
34. _____	**34.** How old is an animal bone that has lost 24% of its carbon-14?
35. _____	**35.** An earthquake was measured at an intensity of $10^{3.91} \cdot I_0$. What was its magnitude on the Richter scale?
36. _____	**36.** The hydrogen ion concentration of a substance is 1.9×10^{-6}. What is the pH?
	37. Solve: $\log_{\sqrt{7}} x = -3$.
37. _____	**38.** Find the domain: $f(x) = \dfrac{1}{\sqrt{3 \ln x - 4}}$.
38. _____	

NAME _____

CLASS _____ SCORE _____ GRADE _____

1. Which of the following relations have inverses that are functions?

a) 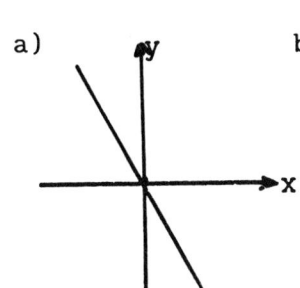   b) 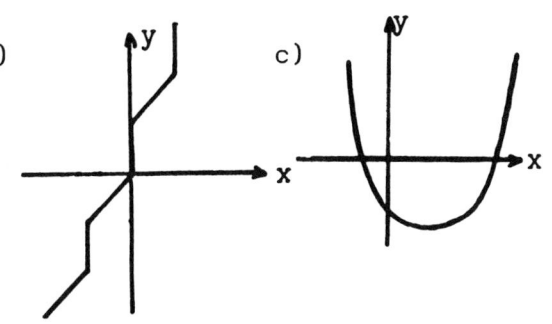   c)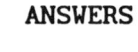

Find a formula for $f^{-1}(x)$.

2. $f(x) = x^2 + 3$

3. $f(x) = 5 + \dfrac{\sqrt{x}}{4}$

4. $h(x) = \dfrac{-5 - 2x}{4}$. Find $h^{-1}(h(6))$.

5. Graph: $y = \log_2 (3x - 1)$.

6. Graph: $f(x) = e^{-3x} - 2$ for nonnegative values of x.

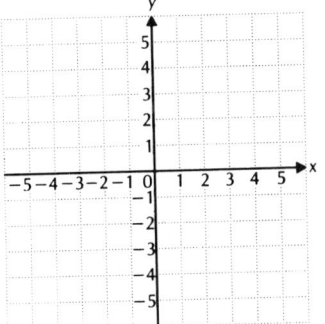   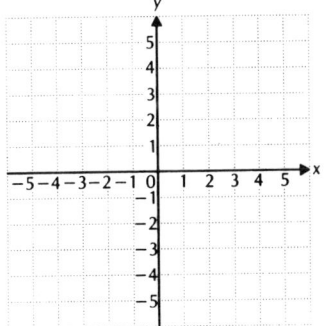

7. Write an exponential equation equivalent to $\log_{\sqrt{8}} 64 = 4$.

8. Write a logarithmic equation equivalent to $a^{-1/3} = 0.00143$.

Simplify.

9. $9^{\log_9 (5x+2)}$

10. $\log_7 7^{x^3}$

11. Find $\log_{12} 58$ using common logarithms.

ANSWERS

1. _____

2. _____

3. _____

4. _____

5. _See graph._

6. _See graph._

7. _____

8. _____

9. _____

10. _____

11. _____

NAME _____

TEST FORM E

ANSWERS

12. _____

13. _____

14. _____

15. _____

16. _____

17. _____

18. _____

19. _____

20. _____

21. _____

22. _____

23. _____

24. _____

25. _____

26. a) _____

    b) _____

27. a) _____

    b) _____

28. _____

29. a) _____

    b) _____

12. Solve for x: $\log_a a^{x^2 - 2x} = 8$.

13. Write an equivalent expression containing a single logarithm.

$$7 \log_a x - \frac{1}{3} \log_a y + \frac{3}{4} \log_a z$$

14. Find the domain: $f(x) = \log_8 (\log_5 x)$.

Solve.

15. $\log_{27} x = -\frac{1}{3}$ 16. $16^x = 64^{5x-2}$

17. $\log_2 (x + 2) + \log_2 (x - 4) = 4$

18. $\log_5 \sqrt{x^2 - 15} = 2$

Find using a calculator.

19. ln 5.14   20. ln 0.000149

21. log (-2.74)   22. log 75,315

Given that $\log_a 2 = 0.301$, $\log_a 12 = 1.079$, and $\log_a 7 = 0.845$, find each of the following.

23. $\log_a \frac{7}{2}$ 24. $\log_a \sqrt{24}$ 25. $\log_a 42$

26. An investment of $5000 doubles in 6 years. What is the interest rate if interest is compounded

    a) annually?   b) continuously?

27. The population of a city was 100,000 in 1960, and the exponential growth rate is 1.5% per year.

    a) What will the population be in 2000?

    b) When will the population be 250,000?

28. How old is an animal bone which has lost 34% of its carbon-14?

29. The average walking speed R of people living in a city of population P, in thousands, is given by
R = 0.37 ln P + 0.05, where R is in feet per second.

    a) The population of San Francisco is 705,300. Find the average walking speed.

    b) A city's population has an average walking speed of 1.7 ft/sec. Find the population.

Write the letter of your response on the answer blank.

ANSWERS

1. Which of the following relations have inverses that are functions?

1. _____

a)

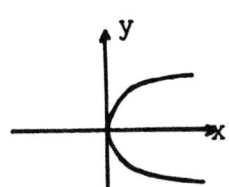

b)

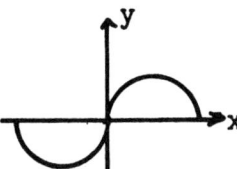

c)

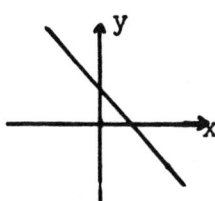

a) a, b   b) a, c   c) b, c   d) b only

2. _____

2. $h(x) = 3(x^2 + 7) - 2x$. Find $h(h^{-1}(9))$.

a) 2   b) 58   c) –9   d) 9

3. _____

3. $f(x) = \sqrt{2x - 3}$. Find a formula for $f^{-1}(x)$.

a) $f^{-1}(x) = 4x^2 + 3$ b) $f^{-1}(x) = 2x - 3$

c) $f^{-1}(x) = \dfrac{x^2 + 3}{2}$ d) $f^{-1}(x) = \sqrt{2x + 3}$

| ANSWERS | 4. Graph $y = \log_2 (x + 1)$. |

4. _____

a)   b)   c)

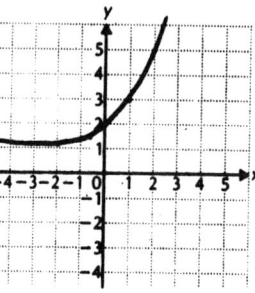   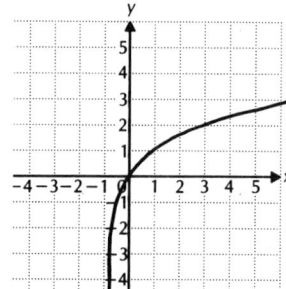   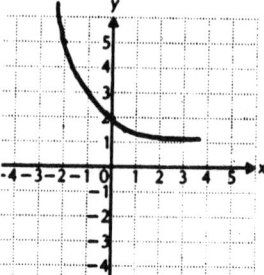

5. _____

5. Graph $f(x) = 1 - e^{-0.5x}$, for nonnegative values of x.

a)   b)   c)

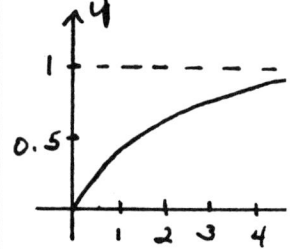   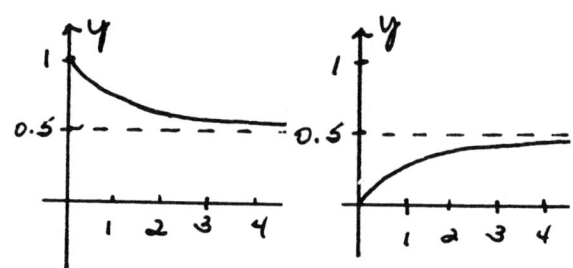

6. _____

6. Find $\log_{0.5} 6$ using natural logarithms.

    a) 1.5563   b) -1.4205   c) -2.5850   d) 0.3802

NAME _____

CLASS _____ SCORE _____ GRADE _____

ANSWERS

7. Write an exponential equation equivalent to $\log_{0.2} 125 = -3$.

7. _____

 a) $125^{0.2} = -3$ b) $10^{0.2} = 125^{-3}$

 c) $125^{-3} = 0.2$ d) $(0.2)^{-3} = 125$

8. Write a logarithmic equation equivalent to $4.5^b = a$.

8. _____

 a) $\log_a b = 4.5$ b) $\log_{4.5} a = b$

 c) $\log_{4.5} b = a$ d) $\log_b 4.5 = a$

9. Write an equivalent expression containing a single logarithm and simplify.

9. _____

$$\log_a 4x + \log_a 3x - \log_a 3 - 2 \log_a 2x$$

 a) $\dfrac{\log_a 6x}{\log_a (3 + x^2)}$ b) $\log_a \dfrac{6x}{3 + x^2}$

10. _____

 c) 0   d) 1

10. Simplify: $\log_6 6^{2x+3}$.

 a) $-\dfrac{3}{2}$ b) 6   c) $2x + 3$ d) $\dfrac{7}{3}$

11. _____

11. Simplify: $7^{\log_7 (x^2-9)}$.

 a) $x^2 - 9$ b) 3   c) 7   d) 7^{x^2-9}

12. Solve for x: $\log_b b^{5x^2-2} = 9x$.

12. _____

 a) -2 b) $\dfrac{1}{5}$ c) $\dfrac{2}{5}$ d) $-\dfrac{1}{5}, 2$

13. Solve: $\log_{49} x = -\dfrac{1}{2}$.

13. _____

 a) 7   b) $\dfrac{1}{7}$ c) $-\dfrac{1}{7}$ d) -24.5

14. Solve: $3^{4x+1} - 5 = 22$.

 a) $-\dfrac{1}{4}$ b) $\dfrac{3}{4}$ c) $\dfrac{1}{2}$ d) -2

14. _____

NAME _____

15. _____

16. _____

17. _____

18. _____

19. _____

20. _____

21. _____

22. _____

23. _____

24. _____

15. Solve: $\log_x (\log_3 27) = 2$.

    a) 3   b) $\frac{3}{2}$ c) $\sqrt{2}$ d) $\sqrt{3}$

16. Solve: $\log_5 (x + 1) - \log_5 x = 2$.

    a) 1   b) $\frac{1}{24}$ c) $\frac{2}{5}$ d) 5

17. If $\log_a x = 2$, $\log_a y = 3$, and $\log_a z = 4$, what is

$$\log_a \frac{\sqrt[3]{xz}}{\sqrt[3]{y^2 z^{-2}}}?$$

    a) $\frac{5}{2}$ b) $\frac{8}{3}$ c) $\frac{10}{3}$ d) $\frac{14}{3}$

18. Using a calculator, find log 0.0000517.

    a) -4.2865   b) -4.7570   c) -4.1457   d) -3.2865

19. Using a calculator, find log 257,301.

    a) 5.3751   b) 13.1755   c) 5.4104   d) 12.4580

20. Using a calculator, find ln 0.00025.

    a) -3.6021   b) -8.2940   c) -3.2840   d) -5.9915

21. How many years will it take an investment of $20,000 to double if interest is compounded continuously at 12.5%? (Find answer to nearest tenth.)

    a) 0.3   b) 4.8   c) 5.3   d) 5.5

22. What is the loudness, in decibels, of a sound whose intensity is 15,000 times I_0?

    a) 42   b) 96   c) 4.2   d) 17,500

23. An earthquake had an intensity of $10^{6.45} \cdot I_0$. What was its magnitude on the Richter scale?

    a) 0.8096   b) 1.8641   c) 6.45   d) 64.5

24. The hydrogen ion concentration of a substance is 3.7×10^{-4}. What is the pH?

    a) 0.5   b) 3.4   c) 5.2   d) 1.9×10^{-4}

NAME _____

CLASS _____ SCORE _____ GRADE _____

Solve.

ANSWERS

1. $0.3x - 0.2y = 1$,
   $0.01x + 0.1y = 0.14$.

2. A boat travels 50 km downstream in 2 hr. It travels 51 km upstream in 3 hr. Find the speed of the boat and the speed of the stream.

3. One week, a business sold 35 sweatshirts. Black ones cost $20.50 and red ones cost $19.75. In all, $707 worth of sweatshirts were sold. How many of each color were sold?

4. The sum of three numbers is 29. The third number is twice the second minus three. The second number is four more than the first. Find the numbers.

Solve using matrices. If there is more than one solution, list three of them.

5. $8x - 4y = 5$,
   $10x + 8y = 3$

6. $2x - y - 2z = 4$,
   $x - 2y + z = -1$,
   $6x - 6y - 2z = 6$

7. $3x - y + z = 2$,
   $x - 2y + z = 1$,
   $4x + y + 2z = 7$

Classify as consistent or inconsistent, dependent or independent.

8. $x - y = -5$,
   $-x - y = 5$

9. $x - 5y + z = 2$,
   $4x + y - z = 3$,
   $3x + 6y - 2z = 1$

10. Find numbers a, b, and c such that the function $f(x) = ax^2 + bx + c$ fits the data points $(0,3)$, $(1,4)$, and $(-2,-5)$. Then write the equation for the function.

Evaluate.

11. $\begin{vmatrix} \frac{1}{2} & 4 \\ -2 & 6 \end{vmatrix}$

12. $\begin{vmatrix} 1 & 3 & -1 \\ 0 & 4 & 1 \\ 1 & 0 & 5 \end{vmatrix}$

1. _____

2. _____

3. _____

4. _____

5. _____

6. _____

7. _____

8. _____

9. _____

10. _____

11. _____

12. _____

NAME _____

13. _____

14. _____

15. _____

16. _____

17. _____

18. _____

19. _____

20. _____

21. _____

22. _____

23. _____

24. _____

25. _____

26. _____

27. _____

Solve using Cramer's rule.

13. $3x + y = 7,$
$2x - 4y = 14$

14. $x + y - z = -4,$
$2x + y + 3z = 9,$
$x + 3y - 2z = -11$

For Exercises 15 - 24, let

$$A = \begin{bmatrix} -1 & 4 \\ 3 & 2 \end{bmatrix} \qquad B = \begin{bmatrix} -1 & 0 & 2 \\ 4 & 2 & 5 \end{bmatrix} \qquad C = \begin{bmatrix} 1 & 0 \\ 2 & -1 \\ 3 & 4 \end{bmatrix}$$

$$D = \begin{bmatrix} 1 & 2 & -3 \\ 4 & 1 & 5 \\ 0 & 2 & 1 \end{bmatrix} \quad E = \begin{bmatrix} 3 & -1 & 2 \\ 0 & 1 & 4 \\ 5 & 2 & 6 \end{bmatrix} \quad F = \begin{bmatrix} 1 & 0 & -4 \end{bmatrix}$$

$$G = \begin{bmatrix} -1 & 5 \\ 2 & -3 \end{bmatrix} \qquad O = \begin{bmatrix} 0 & 0 \\ 0 & 0 \end{bmatrix} \qquad I = \begin{bmatrix} 1 & 0 \\ 0 & 1 \end{bmatrix}.$$

Find each of the following, if possible.

15. GC

16. O + A

17. A + I

18. D + E

19. A - G

20. BC

21. B + C

22. -F

23. 2D - E

24. GI

Find A^{-1}, if it exists.

25. $A = \begin{bmatrix} 3 & 2 \\ -1 & 1 \end{bmatrix}$

26. $A = \begin{bmatrix} -1 & 2 & 1 \\ 3 & -4 & 2 \\ 2 & -4 & -2 \end{bmatrix}$

27. $A = \begin{bmatrix} 2 & -1 & 0 \\ 3 & -2 & 1 \\ 4 & 1 & -3 \end{bmatrix}$

28. Let $A = \begin{bmatrix} 5 & 2 & -1 \\ 3 & 4 & 2 \\ -1 & 3 & 0 \end{bmatrix}$. Find a_{12}, M_{12}, and A_{12}.

28. _____

Evaluate.

29. $\begin{vmatrix} -1 & 2 & 1 \\ 3 & -1 & 2 \\ 1 & 2 & 1 \end{vmatrix}$　　　　　　30. $\begin{vmatrix} -3 & 2 & -2 \\ 1 & 1 & 1 \\ 6 & -4 & 4 \end{vmatrix}$

29. _____

31. $\begin{vmatrix} 2 & 1 & -2 & 3 \\ -1 & -3 & 2 & -1 \\ 2 & 0 & 0 & 0 \\ 3 & 4 & -2 & 1 \end{vmatrix}$　　32. $\begin{vmatrix} -2 & 2 & -2 \\ 3 & 1 & 2 \\ 5 & 1 & 0 \end{vmatrix}$

30. _____

33. Factor: $\begin{vmatrix} bc & ac & ab \\ a & b & c \\ 1 & 1 & 1 \end{vmatrix}$.

31. _____

34. Write a matrix equation equivalent to this system of equations and use the inverse of the coefficient matrix to solve the system. Show all your work.

$$3x - 2y = -11,$$
$$x + 3y = 11$$

32. _____

35. Graph: $2x - 4y \leq 8$.

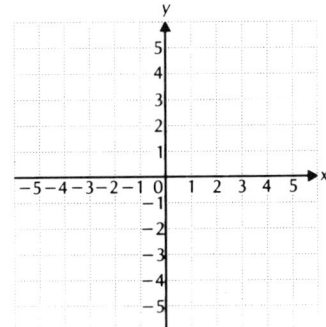

33. _____

34. _____

35. See graph.

NAME _____

36. _____

37. _____

38. _____

39. _____

36. Maximize and minimize $T = 10x + 5y$ subject to

$$x + 3y \leq 15,$$
$$2 \leq x \leq 4,$$
$$y \geq 0.$$

37. A bakery can produce either bread or cake. In a given week, the bakery can turn out at most 700 items, of which 300 loaves of bread and 100 cakes are required by regular customers. The profit on a loaf of bread is $2 and on a cake is $3. How many of each should the bakery produce in order to maximize the profit? What is the maximum profit?

38. Decompose into partial fractions:

$$\frac{-x - 18}{3x^2 - 13x + 4}.$$

39. Solve. $\dfrac{5}{x} - \dfrac{4}{y} = 11,$

$\dfrac{3}{x} + \dfrac{2}{y} = -11.$

NAME _____

CLASS _____ SCORE _____ GRADE _____

Solve.

ANSWERS

1. $0.6x + 0.2y = -0.2$,
   $0.1x + 0.01y = 0.06$.

2. An airplane travels 2700 km with a tail wind in 3 hr. It travels 2600 km with a head wind in 4 hr. Find the speed of the plane and the speed of the wind.

3. Solution A is 10% hydrochloric acid. Solution B is 35% hydrochloric acid. How many liters of each should be mixed to get 60 liters of a solution that is 25% acid?

4. In triangle ABC, the measure of angle B is 8° less than the measure of angle A. The sum of the measures of angles A and B is 84° more than the measure of angle C. Find the angle measures.

Solve using matrices. If there is more than one solution, list three of them.

5. $6x + 4y = 2$,
   $3x + 6y = -1$

6. $2x + y + z = 5$,
   $x - 3y + z = -4$,
   $-x + 2y - z = 2$

7. $4x - 3y + z = 2$,
   $-x + y - z = 4$,
   $5x - 4y + 2z = -2$

Classify as consistent or inconsistent, dependent or independent.

8. $3x - 2y = 0$,
   $-3x + 2y = 7$

9. $x + y - z = 4$,
   $2x - y + z = 2$,
   $x - 3y + 2z = -5$

10. Find numbers a, b, and c such that the function $f(x) = ax^2 + bx + c$ fits the data points $(0,1)$, $(-1,0)$, and $(2,-9)$. Then write the equation for the function.

Evaluate.

11. $\begin{vmatrix} \frac{2}{3} & 1 \\ 5 & -6 \end{vmatrix}$

12. $\begin{vmatrix} 3 & -1 & 2 \\ 0 & 5 & 1 \\ 1 & 2 & 3 \end{vmatrix}$

1. _____

2. _____

3. _____

4. _____

5. _____

6. _____

7. _____

8. _____

9. _____

10. _____

11. _____

12. _____

TEST FORM B

ANSWERS	

13. _____

14. _____

15. _____

16. _____

17. _____

18. _____

19. _____

20. _____

21. _____

22. _____

23. _____

24. _____

25. _____

26. _____

27. _____

Solve using Cramer's rule.

13. $2x - 3y = -22,$
$\;\;\;-3x + 2y = 23$

14. $3x - y + 2z = 7,$
$\;\;\;5x + 2y - z = -8,$
$\;\;\;x + 3y + z = -7$

For Exercises 15 - 24, let

$$A = \begin{bmatrix} 3 & -2 \\ 4 & 1 \end{bmatrix} \quad B = \begin{bmatrix} 2 & -1 & 3 \\ -3 & 4 & 1 \end{bmatrix} \quad C = \begin{bmatrix} 2 & 0 \\ 3 & 1 \\ -1 & 4 \end{bmatrix}$$

$$D = \begin{bmatrix} 1 & 3 & -2 \\ 3 & -1 & 2 \\ 0 & 5 & 1 \end{bmatrix} \quad E = \begin{bmatrix} -2 & 4 & -1 \\ 3 & 0 & -2 \\ -1 & 4 & 1 \end{bmatrix} \quad F = \begin{bmatrix} 2 & 1 & -3 \end{bmatrix}$$

$$G = \begin{bmatrix} -2 & 4 \\ -1 & 3 \end{bmatrix} \quad O = \begin{bmatrix} 0 & 0 \\ 0 & 0 \end{bmatrix} \quad I = \begin{bmatrix} 1 & 0 \\ 0 & 1 \end{bmatrix}.$$

Find each of the following, if possible.

15. GC   16. O + A   17. A + I

18. D + E   19. A − G   20. BC

21. B + C   22. −F   23. 2D − E

24. GI

Find A^{-1}, if it exists.

25. $A = \begin{bmatrix} 2 & -1 \\ 3 & 2 \end{bmatrix}$ 26. $A = \begin{bmatrix} 1 & 4 & 0 \\ 0 & -2 & 3 \\ 2 & -1 & 1 \end{bmatrix}$

27. $A = \begin{bmatrix} 3 & 4 & 1 \\ -9 & 5 & -3 \\ 6 & -1 & 2 \end{bmatrix}$

ANSWERS

28. Let $A = \begin{bmatrix} 3 & -2 & 1 \\ 4 & 0 & -1 \\ 1 & -2 & 3 \end{bmatrix}$. Find a_{23}, M_{23}, and A_{23}.

28. _____

Evaluate.

29. $\begin{vmatrix} 2 & 0 & 5 \\ 1 & -3 & 2 \\ 4 & 1 & 6 \end{vmatrix}$

30. $\begin{vmatrix} 2 & 1 & 3 \\ 1 & 1 & 1 \\ -4 & -2 & 6 \end{vmatrix}$

29. _____

31. $\begin{vmatrix} 2 & -1 & -2 & 3 \\ 5 & 0 & 0 & 0 \\ 4 & 3 & 0 & 1 \\ -6 & 2 & 0 & -1 \end{vmatrix}$

32. $\begin{vmatrix} 3 & -3 & 2 \\ 1 & 0 & 5 \\ -1 & 2 & 4 \end{vmatrix}$

30. _____

33. Factor: $\begin{vmatrix} 1 & 1 & 1 \\ x & y & z \\ x^2 & y^2 & z^2 \end{vmatrix}$.

31. _____

34. Write a matrix equation equivalent to this system of equations and use the inverse of the coefficient matrix to solve the system. Show all your work.

$$3x - y = 9,$$
$$x + 2y = -4$$

32. _____

35. Graph: $3x - 2y \geq 6$.

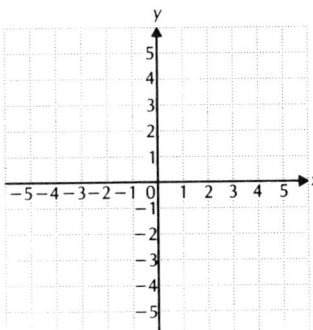

33. _____

34. _____

35.   See graph.

NAME _____

36. _____

37. _____

38. _____

39. _____

36. Maximize and minimize T = 3x + 8y subject to

$$x + 2y \leq 6,$$
$$-1 \leq x \leq 3,$$
$$y \geq 0.$$

37. You are about to take a test which contains questions of Type A worth 10 points each and Type B worth 5 points each. You must complete the test in 60 minutes. Type A questions take 4 minutes and type B questions take 6 minutes. The total number of problems worked must <u>not</u> exceed 10. If you are told that you must work at least 3 questions of Type B, how many of each type of question must you do to maximize your score? What is the maximum score?

38. Decompose into partial fractions:

$$\frac{-5x + 21}{4x^2 + 17x - 15}.$$

39. Two solutions of the equation y = mx + b are (-4,-5) and (2,-2). Find m and b.

NAME _____

CLASS _____ SCORE _____ GRADE _____

Solve.

ANSWERS

1. $0.4x - 0.7y = 1.5,$
$0.03x + 0.1y = -0.04.$

1. _____

2. Two cars leave town traveling in opposite directions. One travels at a speed of 65 km/h and the other at 76 km/h. In how many hours will they be 705 km apart?

2. _____

3. Bill is half as old as his sister Ann. Ten years from now, Bill would be $\frac{3}{4}$ as old as Ann. How old are they now?

3. _____

4. _____

4. A person receives $143 per year in simple interest from three investments totaling $1920. Part is invested at 5%, part at 8%, and part at 10%. There is $420 more invested at 8% than at 5%. Find the amount invested at each rate.

5. _____

Solve using matrices. If there is more than one solution, list three of them.

6. _____

5. $3x - 8y = -3,$
$6x + 4y = -1$

6. $4x + 3y - z = 5,$
$x - 2y + 3z = 4,$
$5x + y + 2z = 9$

7. _____

7. $3x - y - z = 0,$
$4x + 2y + z = 9,$
$x - 2y + z = -2$

8. _____

Classify as consistent or inconsistent, dependent or independent.

9. _____

8. $5x - y = 4,$
$-10x + 2y = -8$

9. $x - y + z = 4,$
$x + 2y + z = 7,$
$-x + y - z = -3$

10. _____

10. Find numbers a, b, and c such that the function $f(x) = ax^2 + bx + c$ fits the data points $(0,1)$, $(2,-3)$, and $(-1,6)$. Then write the equation for the function.

11. _____

Evaluate.

12. _____

11. $\begin{vmatrix} 2 & -\frac{1}{2} \\ 4 & 3 \end{vmatrix}$

12. $\begin{vmatrix} 2 & 1 & 4 \\ 0 & 3 & 2 \\ 1 & 5 & 1 \end{vmatrix}$

NAME _____

ANSWERS	

Solve using Cramer's rule.

13. $5x - 3y = 4,$
    $2x + 4y = -14$

14. $5x + 2y + 3z = -5,$
    $x - y + z = -2,$
    $2x + 3y + 4z = 3$

13. _____

14. _____

For Exercises 15 - 24, let

$$A = \begin{bmatrix} -3 & 1 \\ 0 & 2 \end{bmatrix} \qquad B = \begin{bmatrix} -2 & 3 & 0 \\ 1 & 2 & 4 \end{bmatrix} \qquad C = \begin{bmatrix} 5 & -2 \\ 2 & 0 \\ 1 & -3 \end{bmatrix}$$

15. _____

16. _____

$$D = \begin{bmatrix} 1 & 0 & -2 \\ 4 & -1 & 5 \\ 2 & -3 & 4 \end{bmatrix} \quad E = \begin{bmatrix} 6 & -2 & 1 \\ 4 & -3 & 5 \\ 0 & 2 & 1 \end{bmatrix} \quad F = \begin{bmatrix} 5 & -2 & 7 \end{bmatrix}$$

17. _____

18. _____

$$G = \begin{bmatrix} -3 & -1 \\ -2 & -4 \end{bmatrix} \qquad O = \begin{bmatrix} 0 & 0 \\ 0 & 0 \end{bmatrix} \qquad I = \begin{bmatrix} 1 & 0 \\ 0 & 1 \end{bmatrix}.$$

19. _____

20. _____

Find each of the following, if possible.

21. _____

15. GC   16. O + A   17. A + I

22. _____

18. D + E   19. A - G   20. BC

23. _____

21. B + C   22. -F   23. 2D - E

24. GI

24. _____

Find A^{-1}, if it exists.

25. _____

25. $A = \begin{bmatrix} -1 & 3 \\ 2 & 4 \end{bmatrix}$ 26. $A = \begin{bmatrix} 1 & 0 & -3 \\ 0 & 2 & -4 \\ -2 & 3 & 0 \end{bmatrix}$

26. _____

27. _____

27. $A = \begin{bmatrix} -1 & 1 & 2 \\ -2 & 3 & 1 \\ 4 & 0 & -3 \end{bmatrix}$

NAME _____

28. Let $A = \begin{bmatrix} 1 & -3 & 2 \\ 4 & 5 & -2 \\ 2 & 0 & 1 \end{bmatrix}$. Find a_{31}, M_{31}, and A_{31}.

28. _____

Evaluate.

29. $\begin{vmatrix} 5 & 1 & 3 \\ 2 & -1 & 4 \\ 3 & 0 & 5 \end{vmatrix}$

30. $\begin{vmatrix} 3 & -1 & 2 \\ 1 & 1 & 1 \\ -6 & 2 & -4 \end{vmatrix}$

29. _____

31. $\begin{vmatrix} -1 & 2 & -2 & 1 \\ 0 & 0 & 0 & 3 \\ 4 & 2 & 1 & -5 \\ 1 & -2 & 3 & 4 \end{vmatrix}$

32. $\begin{vmatrix} 8 & 0 & 1 \\ -3 & 5 & 7 \\ 0 & 2 & 4 \end{vmatrix}$

30. _____

33. Factor: $\begin{vmatrix} x & y & z \\ x^2 & y^2 & z^2 \\ 1 & 1 & 1 \end{vmatrix}$.

31. _____

34. Write a matrix equation equivalent to this system of equations and use the inverse of the coefficient matrix to solve the system. Show all your work.

$$2x - 4y = 6,$$
$$3x + 2y = 1$$

32. _____

35. Graph: $x - 5y \leq 5$.

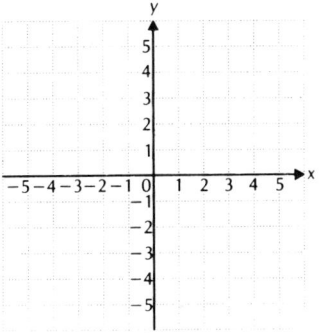

33. _____

34. _____

35. See graph.

NAME _____

36. _____

37. _____

38. _____

39. _____

36. Maximize and minimize $T = 8x + 12y$ subject to

$$x + y \leq 6,$$
$$x \geq 0,$$
$$-1 \leq y \leq 4.$$

37. A tailoring firm can produce either slacks or skirts. In a given week, the firm can turn out at most 200 items, of which 40 pairs of slacks and 50 skirts are required by regular customers. The profit on a pair of slacks is $12 and on a skirt is $15. How many of each should the firm produce in order to maximize profit? What is the maximum profit?

38. Decompose into partial fractions:

$$\frac{-x - 23}{2x^2 - 7x - 4}.$$

39. Solve: $w + x - y + z = 2,$
$w - x + y + z = 6,$
$w + 2x - y - z = -6,$
$w + x + y - z = -4.$

NAME _____

Solve.

ANSWERS

1. $0.4x + 0.5y = -0.7$,
   $0.02x + 0.1y = 0.04$.

2. A train leaves a station and travels north at a speed
   of 90 km/h. Three hours later, a second train leaves
   on a parallel track and travels north at 120 km/h. How
   far from the station will they meet?

3. A collection of 27 coins consists of dimes and nickels.
   The total value is $1.90. How many dimes and how many
   nickels are there?

4. When machines A, B, C are all working, they can produce
   1040 gadgets in one day. When only A and C are
   working, 770 gadgets can be produced in one day. When
   only B and C are working, 690 gadgets can be produced
   in one day. How many gadgets can be produced in a day
   by each machine?

Solve using matrices. If there is more than one solution,
list three of them.

5. $12x - 4y = -11$,
   $8x + 10y = -1$

6. $5x - y - 2z = 0$,
   $x - 2y + z = 6$,
   $x - y - z = -1$

7. $3x + y + z = 0$,
   $x - 2y + z = 4$,
   $x + 5y - z = -8$

Classify as consistent or inconsistent, dependent or
independent.

8. $3x - 2y = 4$,
   $-3x - 2y = 4$

9. $x - 2y + z = 4$,
   $2x - y - z = 5$,
   $x + y - 2z = 1$

10. Find numbers a, b, and c such that the function
    $f(x) = ax^2 + bx + c$ fits the data points $(0,-3)$,
    $(1,-2)$, and $(-2,7)$. Then write the equation for the
    function.

Evaluate.

11. $\begin{vmatrix} \frac{3}{4} & -2 \\ 3 & 8 \end{vmatrix}$

12. $\begin{vmatrix} 5 & 1 & 3 \\ 0 & 2 & 1 \\ 3 & 1 & 4 \end{vmatrix}$

1. _____

2. _____

3. _____

4. _____

5. _____

6. _____

7. _____

8. _____

9. _____

10. _____

11. _____

12. _____

119

NAME _____

ANSWERS

Solve using Cramer's rule.

13. _____

14. _____

13. $2x - 4y = 22,$
    $3x + 7y = 7$

·14. $x - 3y + z = 4,$
    $3x - y + 2z = 13,$
    $2x + 2y - z = 11$

For Exercises 15 - 24, let

15. _____

16. _____

$$A = \begin{bmatrix} -2 & -3 \\ 1 & 4 \end{bmatrix} \quad B = \begin{bmatrix} 3 & -1 & 2 \\ 0 & 2 & 5 \end{bmatrix} \quad C = \begin{bmatrix} 5 & -1 \\ 1 & 3 \\ 0 & 4 \end{bmatrix}$$

17. _____

18. _____

$$D = \begin{bmatrix} 0 & 1 & 3 \\ -2 & 1 & 4 \\ 7 & 0 & 2 \end{bmatrix} \quad E = \begin{bmatrix} -2 & 3 & 1 \\ 5 & 0 & -4 \\ 3 & -1 & 5 \end{bmatrix} \quad F = \begin{bmatrix} -2 & 3 & 5 \end{bmatrix}$$

19. _____

$$G = \begin{bmatrix} -2 & -3 \\ 4 & 6 \end{bmatrix} \quad O = \begin{bmatrix} 0 & 0 \\ 0 & 0 \end{bmatrix} \quad I = \begin{bmatrix} 1 & 0 \\ 0 & 1 \end{bmatrix}.$$

20. _____

Find each of the following, if possible.

21. _____

15. GC   16. O + A   17. A + I

22. _____

18. D + E   19. A - G   20. BC

21. B + C   22. -F   23. 2D - E

23. _____

24. GI

24. _____

Find A^{-1}, if it exists.

25. _____

25. $A = \begin{bmatrix} -4 & 1 \\ 2 & 3 \end{bmatrix}$ 26. $A = \begin{bmatrix} 1 & 1 & -1 \\ -1 & 2 & 3 \\ -2 & -1 & 1 \end{bmatrix}$

26. _____

27. _____

27. $A = \begin{bmatrix} 5 & 0 & -1 \\ -3 & 0 & 4 \\ 2 & 0 & 1 \end{bmatrix}$

TEST FORM D

28. Let $A = \begin{bmatrix} 4 & 0 & -2 \\ 3 & 1 & 5 \\ 2 & -3 & 6 \end{bmatrix}$. Find a_{23}, M_{23}, and A_{23}.

Evaluate.

29. $\begin{vmatrix} 3 & -1 & 3 \\ 0 & 2 & 1 \\ 5 & 6 & -4 \end{vmatrix}$

30. $\begin{vmatrix} 5 & -1 & 3 \\ 1 & 1 & 1 \\ -10 & 2 & -6 \end{vmatrix}$

31. $\begin{vmatrix} 1 & 3 & 0 & 2 \\ 1 & 0 & -2 & 1 \\ -1 & 0 & 1 & -1 \\ 1 & 0 & 4 & 0 \end{vmatrix}$

32. $\begin{vmatrix} 2 & 6 & 2 \\ 3 & 5 & 4 \\ 0 & -3 & -1 \end{vmatrix}$

33. Factor: $\begin{vmatrix} 1 & a^2 & a \\ 1 & b^2 & b \\ 1 & c^2 & c \end{vmatrix}$.

34. Write a matrix equation equivalent to this system of equations and use the inverse of the coefficient matrix to solve the system. Show all your work.

$$5x - y = 11,$$
$$2x + 4y = 22$$

35. Graph: $2x - 3y \geq -6$.

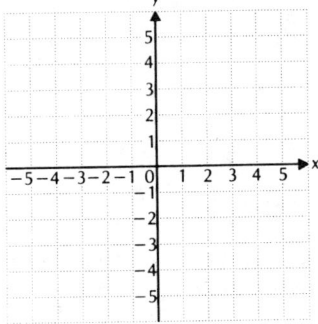

ANSWERS

28. _____

29. _____

30. _____

31. _____

32. _____

33. _____

34. _____

35. See graph.

NAME

ANSWERS

36. _____

37. _____

38. _____

39. _____

36. Maximize and minimize T = 15x + 2y subject to

$$3x + 4y \le 12,$$
$$x \ge -2,$$
$$-3 \le y \le 3.$$

37. A college snack bar cooks and sells hamburgers and hot dogs during the lunch hour. To stay in business, it must sell at least 20 hamburgers but cannot sell more than 60. It must also sell at least 15 hot dogs but cannot cook more than 80. It cannot cook more than 100 sandwiches altogether. The profit on a hamburger is $0.75 and on a hot dog is $0.50. How many of each kind of sandwich should they sell in order to make the maximum profit? What is the maximum profit?

38. Decompose into partial fractions:

$$\frac{-3x + 34}{5x^2 + 17x - 12}.$$

39. Solve: $\begin{vmatrix} x & 2 \\ -8 & x \end{vmatrix} = 20.$

NAME _____

CLASS _____ SCORE _____ GRADE _____

Solve.

ANSWERS

1. $7x - 3y = 5,$
   $6x + 4y = 1$

2. $0.4x + 0.1y = 1.4,$
   $0.3x + 0.4y = -0.9$

1. _____

3. Two investments are made totaling $15,000. For a certain year these investments yield $1095 in simple interest. Part of the $15,000 is invested at 6% and the rest at 9%. Determine the amount invested at each rate.

2. _____

3. _____

4. In a factory there are three machines A, B, and C. When all three are running, they produce 915 skateboards per day. If A and C work but B does not, they produce 695 skateboards per day. If A and B work but C does not, they produce 570 skateboards per day. How many skateboards per day can machine A produce?

4. _____

Solve using matrices. If there is more than one solution, list three of them.

5. _____

5. $x - 2y = 0,$
   $-2x + 4y = 0$

6. $4x + 3y - z = -3,$
   $x - y + 5z = 8,$
   $3x + 4y - 2z = -7$

6. _____

7. Classify as consistent or inconsistent, dependent or independent.

$$2x + y - z = 0,$$
$$-4x - 3y = 3,$$
$$5x + 4y - z = -3$$

7. _____

8. Find numbers a, b, and c such that the function $f(x) = ax^2 + bx + c$ fits the data points $(0,-4)$, $(-1,-11)$, and $(2,-2)$. Then write the equation for the function.

8. _____

NAME _____

9. _____

9. Evaluate: $\begin{vmatrix} -\sqrt{3} & -1 \\ -4 & \sqrt{3} \end{vmatrix}$.

10. _____

10. Solve using Cramer's Rule. Show all your work.

$$2x - z = 2,$$
$$x - 3y + z = 10,$$
$$-3x + y + 4z = 6$$

11. _____

Evaluate.

11. $\begin{vmatrix} 6 & 1 & -2 \\ -6 & 3 & -4 \\ 6 & 1 & 11 \end{vmatrix}$

12. _____

12. $\begin{vmatrix} 3 & 5 & -2 \\ 1 & 0 & -5 \\ 4 & -1 & -1 \end{vmatrix}$

13. _____

13. $\begin{vmatrix} 2 & 6 & 5 \\ 0 & 0 & 0 \\ -7 & -4 & -1 \end{vmatrix}$

14. $\begin{vmatrix} 5 & 0 & 2 & -3 \\ 0 & 9 & -1 & 4 \\ 0 & 0 & -2 & 8 \\ 0 & 0 & 0 & -3 \end{vmatrix}$

14. _____

15. Factor: $\begin{vmatrix} x & 1 & yz \\ y & 1 & xz \\ z & 1 & xy \end{vmatrix}$.

15. _____

16. _____

16. Write a matrix equation equivalent to this system of equations and use the inverse of the coefficient matrix to solve the system. Show all your work.

$$-3x + 4y = 10,$$
$$2x - 5y = -9$$

TEST FORM E

Let

$$A = \begin{bmatrix} -1 & 0 & 2 \\ 0 & 2 & -3 \\ 4 & 0 & -2 \end{bmatrix} \quad B = \begin{bmatrix} 1 & 0 \\ -4 & 2 \\ -2 & -3 \end{bmatrix} \quad C = \begin{bmatrix} 1 & 4 & -3 \\ 0 & -10 & 2 \end{bmatrix}$$

$$D = \begin{bmatrix} 0 & 4 \\ 5 & -2 \end{bmatrix} \quad E = \begin{bmatrix} 1 & 5 & 4 \\ 0 & -2 & 0 \\ -1 & -1 & 1 \end{bmatrix} \quad F = \begin{bmatrix} 3 & -1 \\ 0 & 0 \end{bmatrix}$$

$$G = \begin{bmatrix} 7 & 0 & -2 \end{bmatrix}$$

Matching:
Find each of the following, if possible. Place the letter of the answer in the answer blank. Some letters may be used more than once, and some may not be used.

Questions

17. D + F

18. FB

19. −G

20. DC

21. A^{-1}

22. E − 2A

23. D^{-1}

Answers

a) $\begin{bmatrix} -7 & 0 & 2 \end{bmatrix}$ b) $\begin{bmatrix} 0 & 16 & -12 \\ 0 & -50 & -4 \end{bmatrix}$

c) $\begin{bmatrix} 1 & 0 \\ 4 & -10 \\ -3 & 2 \end{bmatrix}$ d) $\begin{bmatrix} 3 & 5 & 0 \\ 0 & -6 & 6 \\ -9 & -1 & 5 \end{bmatrix}$

e) $\begin{bmatrix} 7 \\ 0 \\ -2 \end{bmatrix}$ f) $\begin{bmatrix} 3 & 3 \\ 5 & -2 \end{bmatrix}$

g) $\begin{bmatrix} \frac{1}{3} & 0 & \frac{1}{3} \\ 1 & \frac{1}{2} & \frac{1}{4} \\ \frac{2}{3} & 0 & \frac{1}{6} \end{bmatrix}$ h) $\begin{bmatrix} \frac{1}{5} & 0 & \frac{1}{5} \\ \frac{3}{5} & \frac{3}{10} & \frac{3}{20} \\ \frac{2}{5} & 0 & \frac{1}{10} \end{bmatrix}$

i) $\begin{bmatrix} 0 & 5 & 6 \\ 0 & 0 & -3 \\ 3 & -1 & -1 \end{bmatrix}$ j) $\begin{bmatrix} 0 & -40 & 8 \\ 5 & 40 & -19 \end{bmatrix}$

k) $\begin{bmatrix} \frac{1}{10} & \frac{1}{5} \\ \frac{1}{4} & 0 \end{bmatrix}$ l) Not possible

m) None of these answers is correct.

ANSWERS

17. _____

18. _____

19. _____

20. _____

21. _____

22. _____

23. _____

NAME _____

ANSWERS	

24. See graph.

24. Graph: $-2x \geq 3y - 6$.

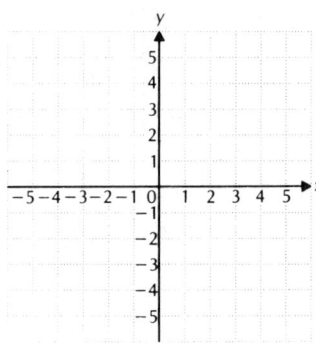

25. _____

25. Maximize and minimize $T = 20x + 15y$ subject to

$$3x + 2y \leq 6,$$
$$x + y \leq 3,$$
$$x \geq 0,$$
$$y \geq 1$$

26. _____

26. It takes a tailoring firm 1 hr of cutting and 4 hr of sewing to make a dress. To make a suit it takes 1 hr of cutting and 6 hr of sewing. At most, 5 hr per day are available for cutting, and at most 24 hr per day are available for sewing. The profit on a dress is $25 and on a suit is $35. How many of each type of garment should be made to maximize profit? What is the maximum profit?

NAME _____

CLASS _____ SCORE _____ GRADE _____

Write the letter of your response on the answer blank.

ANSWERS

1. Solve. $4x + 5y = -2,$
   $3x + 4y = -1$

   a) $\left(-2, \frac{6}{5}\right)$ or $(-3, 2)$ b) $\left(\frac{5}{7}, -\frac{2}{7}\right)$ or $\left(1, -\frac{6}{5}\right)$

   c) $(3, 2)$ or $(-5, 7)$ d) $(6, -6)$ or $(4, -5)$

1. _____

2. Solve: $\dfrac{6}{x} + \dfrac{2}{y} = -2,$

   $\dfrac{7}{x} - \dfrac{3}{y} = -13.$

   The solution set

   a) is empty.

   b) contains one ordered pair.

   c) contains four ordered pairs.

   d) contains two ordered pairs.

2. _____

3. The sum of a certain number and a second number is 15.
   The second number minus the first number is -119.
   Find the smaller number.

   a) 67   b) -52   c) 52   d) -67

3. _____

4. Two investments are made totaling $19,000. For a
   certain year these investments yield $2520 in simple
   interest. Part of the $19,000 is invested at 12% and
   the rest at 14%. How much is invested at 12%?

   a) $12,000   b) $8,500   c) $9,000   d) $7000

4. _____

5. Solve using matrices: $2x + 3y - 5z = -10,$
   $-3x + 2y + z = 2,$
   $5x - y + 3z = 13.$

   Find the sum of the x, y, and z values.

   a) 0   b) -4   c) -1   d) 5

5. _____

6. _____

7. _____

8. _____

9. _____

10. _____

6. Classify this system as consistent or inconsistent, dependent or independent.

$$2x - 5y = 4,$$
$$-6x + 15y = 6$$

a) Consistent, dependent

b) Consistent, independent

c) Inconsistent, dependent

d) Inconsistent, independent

7. Classify this system as consistent or inconsistent, dependent or independent.

$$4x + z = 5,$$
$$x + y + z = -2,$$
$$5x + y + z = 3$$

a) Consistent, dependent

b) Consistent, independent

c) Inconsistent, dependent

d) Inconsistent, independent

8. Find numbers a, b, and c such that the function $f(x) = ax^2 + bx + c$ fits the data points $(0,1)$, $(-1,9)$, and $(2,3)$.

a) $a = -1$, $b = 4$, $c = 1$ b) $a = 1$, $b = -2$, $c = -5$

c) $a = 3$, $b = -5$, $c = 1$ d) $a = 3$, $b = 1$, $c = -3$

9. Evaluate: $\begin{vmatrix} -\dfrac{1}{4} & -\sqrt{2} \\ \sqrt{2} & 8 \end{vmatrix}$.

a) -4 b) -6 c) 2 d) 0

10. Evaluate: $\begin{vmatrix} 4 & 2 & -3 \\ -2 & -3 & 0 \\ 1 & 2 & 2 \end{vmatrix}$.

a) 14 b) 35 c) -13 d) -37

11. Solve using Cramer's Rule: 2x − 3y = 13,
                                  9x + 5y = 9.

   Find the y-coordinate of the solution.

   a) $\dfrac{\begin{vmatrix} 2 & 13 \\ 9 & 9 \end{vmatrix}}{\begin{vmatrix} 2 & -3 \\ 9 & 5 \end{vmatrix}}$ b) $\dfrac{\begin{vmatrix} 13 & 13 \\ 9 & 9 \end{vmatrix}}{\begin{vmatrix} 2 & -3 \\ 9 & 5 \end{vmatrix}}$ c) $\dfrac{\begin{vmatrix} 13 & 9 \\ 9 & 5 \end{vmatrix}}{\begin{vmatrix} 2 & -3 \\ 9 & 5 \end{vmatrix}}$ d) $\dfrac{\begin{vmatrix} 2 & -3 \\ 9 & 5 \end{vmatrix}}{\begin{vmatrix} 2 & -3 \\ 9 & 5 \end{vmatrix}}$

12. Solve using Cramer's Rule: 7x − y + z = −3,
                                 x + 3y + 2z = 2,
                                 −3x − 2y − z = 2.

   Find the sum of the x, y, and z values.

   a) −1   b) 2   c) −4   d) 1

Let

$A = \begin{bmatrix} 4 & 7 & 0 \\ -2 & 0 & 3 \\ 0 & 1 & -2 \end{bmatrix}$ $B = \begin{bmatrix} 2 & -2 & 3 \\ 0 & 1 & -4 \\ -1 & -3 & 2 \end{bmatrix}$ $C = \begin{bmatrix} 0 & 2 \\ -2 & 1 \\ 3 & -1 \end{bmatrix}$

$D = \begin{bmatrix} -1 & 0 & -2 \\ 4 & 0 & -2 \end{bmatrix}$ $E = \begin{bmatrix} -2 & -2 \\ 4 & 4 \end{bmatrix}$ $F = \begin{bmatrix} 0 & 2 \\ -3 & 1 \end{bmatrix}$

$G = \begin{bmatrix} 5 \\ 0 \\ -2 \end{bmatrix}$ $H = \begin{bmatrix} 2 & -8 & 4 \end{bmatrix}$ $I = \begin{bmatrix} 1 & 0 & 0 \\ 0 & 1 & 0 \\ 0 & 0 & 1 \end{bmatrix}$

$O = \begin{bmatrix} 0 & 0 \\ 0 & 0 \end{bmatrix}$

Find each of the following (Questions 13 − 24), if possible.

13. CD

   a) $\begin{bmatrix} 0 & 0 & 2 \\ -20 & -1 & 2 \end{bmatrix}$ b) $\begin{bmatrix} 8 & 0 & -4 \\ 6 & 0 & 2 \\ -7 & 0 & -4 \end{bmatrix}$

   c) $\begin{bmatrix} -20 & -1 & 8 \end{bmatrix}$ d) $\begin{bmatrix} 20 & 2 & 4 \\ 6 & 1 & -2 \\ -7 & -1 & 4 \end{bmatrix}$

ANSWERS

11. _____

12. _____

13. _____

ANSWERS	14. 3D

14. _____

a) $\begin{bmatrix} 3 & 0 & 6 \\ 4 & 0 & -2 \end{bmatrix}$ b) $\begin{bmatrix} -3 & 0 & -2 \\ 12 & 0 & -2 \end{bmatrix}$

c) $\begin{bmatrix} -3 & 0 & -6 \\ 12 & 0 & -6 \end{bmatrix}$ d) $\begin{bmatrix} 4 & 3 & 5 \\ -7 & 2 & 1 \end{bmatrix}$

15. G + H

15. _____

a) $\begin{bmatrix} 3 \\ -8 \\ 2 \end{bmatrix}$ b) $\begin{bmatrix} 7 & -8 & 2 \end{bmatrix}$

c) $\begin{bmatrix} 3 & -8 & 2 \end{bmatrix}$

d) Not possible

16. E − F

a) $\begin{bmatrix} -2 & -4 \\ 7 & 3 \end{bmatrix}$ b) $\begin{bmatrix} -2 & -4 \\ 1 & 3 \end{bmatrix}$ c) $\begin{bmatrix} 6 & -6 \\ -12 & 12 \end{bmatrix}$ d) $\begin{bmatrix} -2 & 0 \\ 1 & 5 \end{bmatrix}$

16. _____

17. DF

a) $\begin{bmatrix} -12 & -8 \\ 3 & -1 \\ 4 & 2 \end{bmatrix}$ b) $\begin{bmatrix} 0 & 0 & 3 \\ -12 & 6 & 2 \end{bmatrix}$

c) $\begin{bmatrix} 6 & 0 & 12 \\ -60 & -6 & -12 \end{bmatrix}$ d) Not possible

17. _____

18. −A

a) $\begin{bmatrix} -3 & 14 & 21 \\ -4 & -8 & -12 \\ -2 & -4 & 14 \end{bmatrix}$ b) $\begin{bmatrix} \frac{3}{40} & -\frac{7}{20} & -\frac{21}{40} \\ \frac{1}{10} & \frac{1}{5} & \frac{3}{10} \\ \frac{1}{20} & \frac{1}{10} & -\frac{7}{20} \end{bmatrix}$

18. _____

c) $\begin{bmatrix} 4 & -2 & 0 \\ 7 & 0 & 1 \\ 0 & 3 & -2 \end{bmatrix}$ d) $\begin{bmatrix} -4 & -7 & 0 \\ 2 & 0 & -3 \\ 0 & -1 & 2 \end{bmatrix}$

NAME _____

19. $10A + 2B$

a) $\begin{bmatrix} 44 & 66 & 6 \\ -20 & 2 & 22 \\ -2 & 4 & -16 \end{bmatrix}$

b) $\begin{bmatrix} 36 & 66 & 6 \\ -20 & 2 & 22 \\ -2 & 4 & -24 \end{bmatrix}$

c) $\begin{bmatrix} 36 & 74 & -6 \\ -20 & -2 & 38 \\ -2 & 16 & -24 \end{bmatrix}$

d) $\begin{bmatrix} 44 & -20 & 2 \\ 74 & -2 & 16 \\ -6 & 38 & -24 \end{bmatrix}$

19. _____

20. E^{-1}

a) $\begin{bmatrix} 2 & 2 \\ -4 & -4 \end{bmatrix}$

b) $\begin{bmatrix} -2 & 4 \\ -2 & 4 \end{bmatrix}$

c) $\begin{bmatrix} -\dfrac{1}{4} & -\dfrac{1}{8} \\ \dfrac{1}{4} & \dfrac{1}{8} \end{bmatrix}$

d) Not possible

20. _____

21. $F + 0$

a) $\begin{bmatrix} 0 & 2 \\ -3 & 1 \end{bmatrix}$

b) $\begin{bmatrix} 0 & 0 \\ 0 & 0 \end{bmatrix}$

c) $\begin{bmatrix} 0 & -3 \\ 2 & 1 \end{bmatrix}$

d) $\begin{bmatrix} 0 & -2 \\ 3 & -1 \end{bmatrix}$

21. _____

22. $B + I$

a) $\begin{bmatrix} 2 & -2 & 3 \\ 0 & 1 & -4 \\ -1 & -3 & 2 \end{bmatrix}$

b) $\begin{bmatrix} 3 & -2 & 3 \\ 0 & 2 & -4 \\ -1 & -3 & 3 \end{bmatrix}$

c) $\begin{bmatrix} -2 & 2 & -3 \\ 0 & -1 & 4 \\ 1 & 3 & -2 \end{bmatrix}$

d) $\begin{bmatrix} 3 & -1 & 4 \\ 1 & 2 & -3 \\ 0 & -2 & 3 \end{bmatrix}$

22. _____

ANSWERS	23. A^{-1}

23. _____

a) $\begin{bmatrix} -3 & -4 & -2 \\ 14 & -8 & -4 \\ 21 & -12 & 14 \end{bmatrix}$

b) $\begin{bmatrix} \dfrac{3}{40} & -\dfrac{7}{20} & -\dfrac{21}{40} \\[6pt] \dfrac{1}{10} & \dfrac{1}{5} & \dfrac{3}{10} \\[6pt] \dfrac{1}{20} & \dfrac{1}{10} & -\dfrac{7}{20} \end{bmatrix}$

c) $\begin{bmatrix} -4 & -7 & 0 \\ 2 & 0 & -3 \\ 0 & -1 & 2 \end{bmatrix}$

d) $\begin{bmatrix} -3 & 14 & 21 \\ -4 & -8 & -12 \\ -2 & -4 & 14 \end{bmatrix}$

24. _____

24. $A + B$

a) $\begin{bmatrix} 8 & -1 & -16 \\ -7 & -5 & 0 \\ 2 & 7 & -8 \end{bmatrix}$

b) $\begin{bmatrix} 6 & 5 & 3 \\ -2 & 1 & -1 \\ -1 & -2 & 0 \end{bmatrix}$

25. _____

c) $\begin{bmatrix} 6 & 5 & 0 \\ 0 & 0 & -1 \\ 0 & -2 & 0 \end{bmatrix}$

d) $\begin{bmatrix} 8 & -14 & 0 \\ 0 & 0 & -12 \\ 0 & -3 & -4 \end{bmatrix}$

25. Evaluate: $\begin{vmatrix} 3 & 4 & 2 \\ -3 & 1 & -1 \\ -3 & 0 & -2 \end{vmatrix}$.

26. _____

a) -48 b) 0 c) -24 d) -12

26. Evaluate: $\begin{vmatrix} 0 & 2 & 4 \\ -1 & 1 & -3 \\ 4 & -5 & 12 \end{vmatrix}$.

a) -84 b) -36 c) 4 d) -84

27. _____

27. Evaluate: $\begin{vmatrix} -2 & 0 & 0 & 0 \\ 4 & 0 & -6 & 1 \\ 1 & 3 & 0 & 2 \\ -2 & 1 & -3 & 0 \end{vmatrix}$.

a) 6 b) 42 c) -42 d) 0

---

28. Factor: $\begin{vmatrix} 1 & c & c^2 \\ 1 & d & d^2 \\ 1 & e & e^2 \end{vmatrix}$.

   a) $(c - d)(e - c)(d - e)$

   b) $(d - c)(e - c)(d - e)$

   c) $(d - c)(e - c)(e - d)$

   d) $(d - c)(c - e)(d - e)$

28. _____

29. Write a matrix equation equivalent to this system of equations.
$$9x - y = 7,$$
$$3x + 5y = -2$$

a) $\begin{bmatrix} 9 & -1 \\ 3 & 5 \end{bmatrix} \begin{bmatrix} x \\ y \end{bmatrix} = \begin{bmatrix} 7 \\ -2 \end{bmatrix}$
   b) $\begin{bmatrix} 9 & -1 \\ 3 & 5 \end{bmatrix} \begin{bmatrix} x & y \end{bmatrix} = \begin{bmatrix} 7 \\ -2 \end{bmatrix}$

c) $\begin{bmatrix} 7 & -1 \\ -2 & 5 \end{bmatrix} \begin{bmatrix} x \\ y \end{bmatrix} = \begin{bmatrix} 9 \\ 3 \end{bmatrix}$
   d) $\begin{bmatrix} 9 & -1 & 7 \\ 3 & 5 & -2 \end{bmatrix} \begin{bmatrix} x \\ y \end{bmatrix} = \begin{bmatrix} 7 \\ -2 \end{bmatrix}$

29. _____

30. Graph: $2x < 4y + 8$.

a)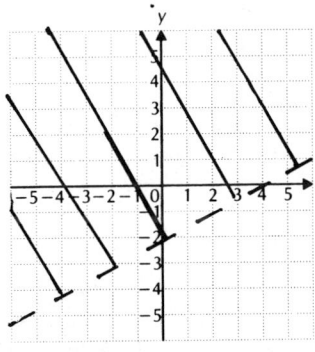
     b)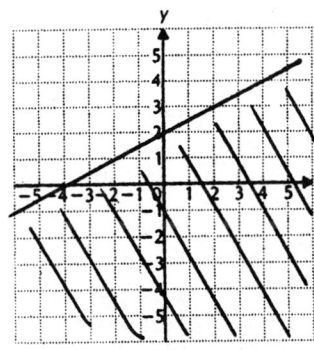

c)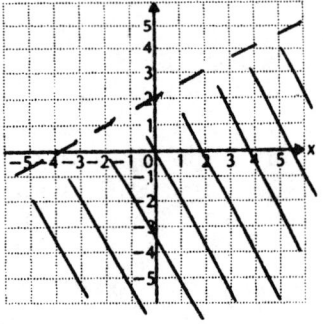
     d)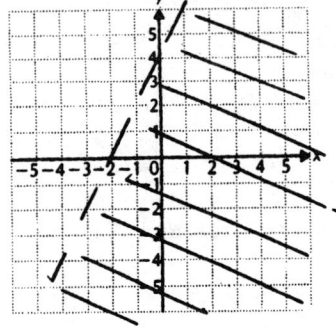

30. _____

NAME _____

31. _____

31. Graph this system: $2x + 11y \leq 22$,
$1 \leq x \leq 8$,
$y \geq 0$.

Find the coordinates of any vertices found.

a) $(1,0)$, $\left(1, \frac{11}{2}\right)$, $(8,0)$, $\left(8, \frac{38}{11}\right)$

b) $(1,0)$, $\left(1, \frac{20}{11}\right)$, $\left(8, \frac{6}{11}\right)$, $(8,0)$

c) $(0,1)$, $\left(\frac{20}{11}, 1\right)$, $\left(\frac{6}{11}, 8\right)$, $(0,8)$

d) $(1,0)$, $(0,2)$, $(8,0)$, $(11,0)$

32. _____

32. Maximize and minimize $P = 11x + 99y$ subject to the system given in Question 31.

a) Min. 11 at $(1,0)$, max. 191 at $\left(1, \frac{20}{11}\right)$

b) Min. 11 at $(1,0)$, max. 342 at $\left(1, \frac{30}{11}\right)$

c) Min. 11 at $(1,0)$, max. 792 at $(0,8)$

d) Min. 99 at $(0,1)$, max. 798 at $\left(\frac{6}{11}, 8\right)$

33. _____

33. You are about to take a test that contains questions of Type A worth 10 points and questions of Type B worth 20 points. You must complete the test in 120 minutes, and the total number of problems worked must not exceed 24. Type A questions take 4 minutes and Type B questions take 12 minutes. If you are told that you must work at least 2 questions of Type B, how many of each type of question must you do to maximize your score and what is the maximum score?

a) Type A: 0,   Type B: 10,   max. Score 200

b) Type A: 12,   Type B: 12,   max. Score 360

c) Type A: 21,   Type B: 3,   max. Score 270

d) Type A: 22,   Type B: 2,   max. Score 260

NAME _____

CLASS _____ *SCORE* _____ *GRADE* _____

1. Graph: $4x^2 + 7xy - 2y^2 = 0$.

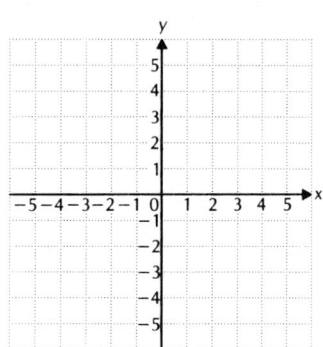

2. Find the center, vertices, and foci of the ellipse $9x^2 + y^2 - 36x - 6y + 36 = 0$, then graph the ellipse.

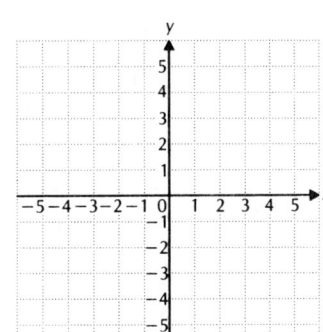

3. Find an equation of the ellipse with vertices $(-2,0)$, $(2,0)$, $(0,-4)$, and $(0,4)$.

4. Find the center, vertices, foci, and asymptotes of the hyperbola $4y^2 - 16x^2 = 64$.

5. Graph: $xy = 2$.

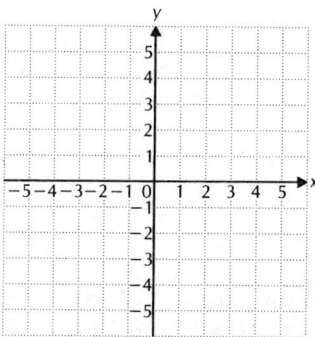

ANSWERS

1. See graph.

2. _____

    See graph.

3. _____

4. _____

5. See graph.

NAME _____

ANSWERS	

6. Find an equation of the parabola with directrix $x = -6$ and focus $(6,0)$.

6. _____

7. _____

7. Find the vertex, focus, and directrix of the parabola $x^2 + 2x - 4y + 13 = 0$.

8. _____

Solve.

8. $x^2 + y^2 = 100$,
   $3x + 4y = 0$

9. $x^2 + y^2 = 2$,
   $xy = 1$

9. _____

10. _____

10. The sum of two numbers is 17, and the sum of their squares is 149. What are the numbers?

Solve.

ANSWERS

11. The diagonal of a rectangle is 3 ft longer than the length of the rectangle and 3 ft shorter than twice the width. Find the dimensions of the rectangle.

11. _____

12. In a fractional expression, the sum of the values of the numerator and the denominator is 21. The product of their values is 110. Find the values of the numerator and the denominator.

12. _____

13. _____

13. The area of a rectangle is 24 ft^2, and the length of a diagonal is $2\sqrt{37}$. Find the dimensions.

14. Find two numbers whose product is 72 if the sum of their squares is 340.

14. _____

15. Two squares are such that the sum of their areas is 34 ft^2 and the difference of their areas is 16 ft^2. Find the length of a side of each square.

15. _____

NAME _____

ANSWERS	

Classify the equation as a circle, an ellipse, a parabola, or a hyperbola.

16. _____

16. $x = -y^2 + 3y - 2$

17. _____

17. $3x^2 - xy = 4 + 3x^2$

18. _____

18. $x^2 + y^2 - 3x + 12y + 29 = 0$

19. _____

19. $12x^2 + 6y^2 = 6 - 48x$

20. _____

20. $y = x^2 + 6x - 2 + 23y$

21. _____

21. $\dfrac{x^2}{25} - \dfrac{y^2}{9} = 1$

22. Find an equation of the ellipse with vertices $(5,-1)$, $(1,-1)$, $(3,-4)$, and $(3,2)$.

22. _____

23. Find an equation of the following parabola: Line of symmetry parallel to the y-axis, vertex $(2,-1)$, and passing through $(0,0)$.

23. _____

24. The square of a certain number exceeds twice the square of another number by $\dfrac{1}{18}$. The sum of their squares is $\dfrac{5}{36}$. Find the numbers.

24. _____

1. Graph: $2x^2 + 5xy - 3y^2 = 0$.

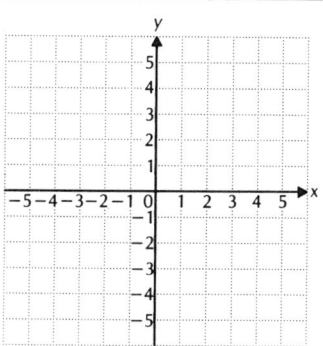

ANSWERS

1. See graph. _____

2. Find the center, vertices, and foci of the ellipse $x^2 + 4y^2 - 2x + 8y - 11 = 0$, then graph the ellipse.

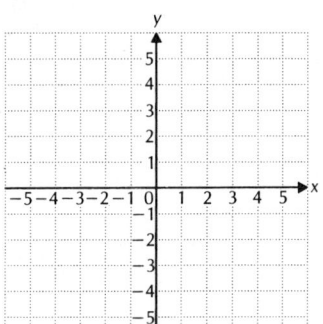

2. _____

See graph. _____

3. Find an equation of the ellipse with vertices $(-5,0)$, $(5,0)$, $(0,-1)$, and $(0,1)$.

3. _____

4. Find the center, vertices, foci, and asymptotes of the hyperbola $16x^2 - 9y^2 = 144$.

4. _____

5. Graph: $xy = -4$.

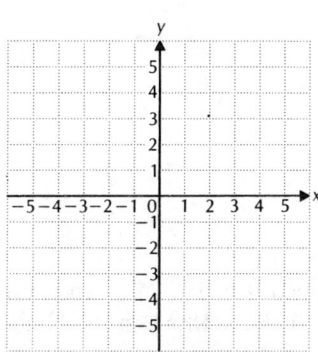

5. See graph. _____

ANSWERS	

6. _____

7. _____

8. _____

9. _____

10. _____

6. Find an equation of the parabola with directrix y = -8 and focus (0,8).

7. Find the vertex, focus, and directrix of the parabola $y^2 + 14y + 4x + 33 = 0$.

Solve.

8. $y^2 - x^2 = 16,$
   $x + y = 4$

9. $x^2 + y^2 = 25,$
   $y^2 + x = 5$

10. The sum of two numbers is 25, and the difference of their squares is 125. What are the numbers?

TEST FORM B

Solve.	ANSWERS
11. A rectangle has a perimeter of 24 cm and the length of a diagonal is $\sqrt{74}$ cm. Find the dimensions.	11. _____
12. In a fractional expression, the sum of the values of the numerator and the denominator is 14. The product of their values is 45. Find the values of the numerator and the denominator.	12. _____
13. A rectangle with diagonal of length $\sqrt{106}$ has an area of 45. Find the dimensions of the rectangle.	13. _____
14. Find two numbers whose product is 66 if the sum of their squares is 157.	14. _____
15. A garden contains two square flower beds. Find the length of each bed if the sum of their areas is 113 ft^2 and the difference of their areas is 15 ft^2.	15. _____

NAME _____

ANSWERS	Classify the equation as a circle, an ellipse, a parabola, or a hyperbola.

16. _____

16. $y^2 + x^2 + 10x - 4y + 28 = 0$

17. _____

17. $25x^2 + 9y^2 + 150x - 18y = -9$

18. _____

18. $y = 2x^2 - 3x + 8$

19. _____

19. $\dfrac{(y - 3)^2}{4} - \dfrac{(x - 5)^2}{16} = 1$

20. _____

20. $x = y^2 - 6y + 17 - 3x$

21. _____

21. $xy - 9y^2 = 18 - 9y^2$

22. Find an equation of a hyperbola having $y = \dfrac{3}{4}x$ and $y = -\dfrac{3}{4}x$ as asymptotes and one vertex $(4,0)$.

22. _____

23. Find two numbers whose product is 10 and the sum of whose reciprocals is $\dfrac{7}{10}$.

23. _____

24. _____

24. Find an equation of a circle that passes through the points $(1,-1)$ and $(2,-2)$ and whose center is on the line $3x - y = 5$.

NAME

CLASS _____ SCORE _____ GRADE _____

1. Graph: $3x^2 + 5xy - 2y^2 = 0$.

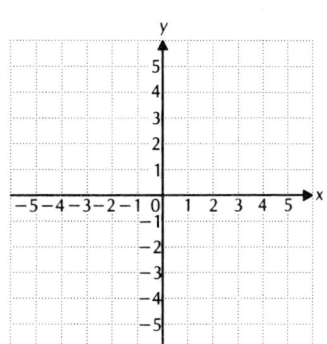

ANSWERS

1. See graph.

2. _____

   See graph.

2. Find the center, vertices, and foci of the ellipse $4x^2 + 9y^2 - 16x + 54y + 61 = 0$, then graph the ellipse.

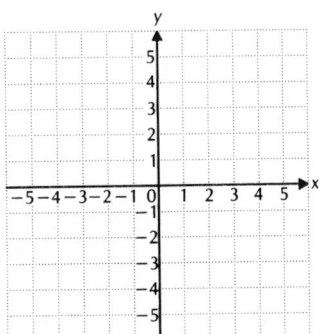

3. _____

3. Find an equation of the ellipse with vertices $(-6,0)$, $(6,0)$, $(0,-5)$, and $(0,5)$.

4. Find the center, vertices, foci, and asymptotes of the hyperbola $9x^2 - 4y^2 = 36$.

4. _____

5. Graph: $xy = 3$.

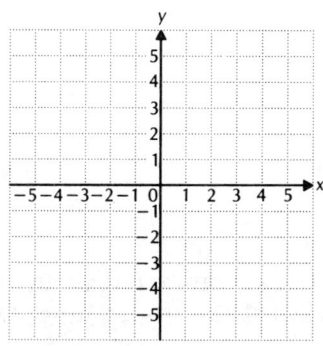

5. See graph.

NAME _____

6. _____

7. _____

8. _____

9. _____

10. _____

6. Find an equation of the parabola with directrix $x = 3$ and focus $(-3,0)$.

7. Find the vertex, focus, and directrix of the parabola $x^2 - 2x + 4y + 21 = 0$.

Solve.

8. $y^2 = x + 5,$
   $3y = x + 7$

9. $2x^2 - y^2 = 4,$
   $x^2 + 2y^2 = 12$

10. The difference of two numbers is 9, and the difference of their squares is 135. What are the numbers?

NAME _____

Solve. | ANSWERS
---

11. A rectangle has a area of 32 in.^{2} and a perimeter of of 24 in. Find the dimensions.

11. _____

12. In a fractional expression, the sum of the values of the numerator and the denominator is 21. The product of their values is 104. Find the values of the numerator and the denominator.

12. _____

13. The area of a rectangle is 60 yd^2, and the length of a diagonal is 13 yd. Find the dimensions.

13. _____

14. Find two numbers whose product is 75 if the sum of their squares is 250.

14. _____

15. Two squares are such that the sum of their areas is 53 cm^2 and the difference of their areas is 45 cm^2. Find the length of a side of each square.

15. _____

NAME _____

ANSWERS	Classify the equation as a circle, an ellipse, a parabola, or a hyperbola.

16. _____

16. $x^2 + y^2 - 18x + 8y + 33 = 0$

17. _____

17. $x = 5 + 3y - y^2$

18. _____

18. $xy + 17 + 9x^2 = 9x^2$

19. _____

19. $36x^2 + 25y^2 = 900$

20. _____

20. $x^2 - 25y^2 + 100y - 125 = 0$

21. _____

21. $x^2 - 46y = 0$

22. Find an equation of the ellipse with vertices $(3,3)$, $(-7,3)$, $(-2,7)$, and $(-2,-1)$.

22. _____

23. Find two numbers whose product is 12 and the sum of whose reciprocals is $\frac{7}{12}$.

23. _____

24. Find an equation of a circle that passes through the points $(-3,3)$ and $(-2,2)$ and whose center is on the line $2x + 5y = 4$.

24. _____

NAME _____

CLASS _____ *SCORE* _____ *GRADE* _____

1. Graph: $5x^2 + 3xy - 2y^2 = 0$.

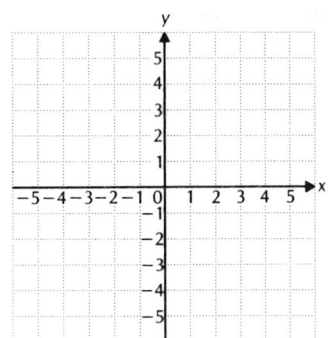

ANSWERS

1. See graph. _____

2. Find the center, vertices, and foci of the ellipse $9x^2 + 16y^2 - 36x - 32y - 92 = 0$, then graph the ellipse.

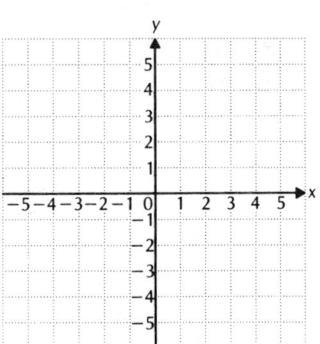

2. _____

   See graph. _____

3. Find an equation of the ellipse with vertices $(-4,0)$, $(4,0)$, $(0,-3)$, and $(0,3)$.

4. Find the center, vertices, foci, and asymptotes of the hyperbola $25y^2 - 9x^2 = 225$.

3. _____

4. _____

5. Graph: $xy = -5$.

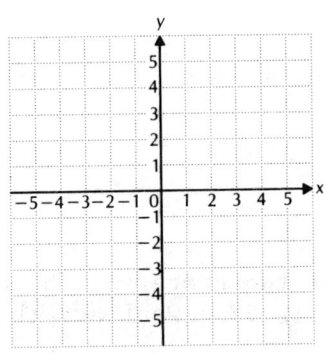

5. See graph. _____

NAME _____

ANSWERS

6. _____

7. _____

8. _____

9. _____

10. _____

6. Find an equation of the parabola with directrix $y = 6$ and focus $(0,-6)$.

7. Find the vertex, focus, and directrix of the parabola $y^2 - 6y - 4x + 17 = 0$.

Solve.

8. $y^2 = x - 4$,
   $3y = x - 2$

9. $x^2 + 9y^2 = 90$,
   $xy = 9$

10. The difference of two numbers is 5, and the sum of their squares is 97. What are the numbers?

NAME _____

Solve.

ANSWERS

11. It will take 310 yd of fencing to enclose a rectangular field. The area of the field is 4600 yd^2. What are the dimensions?

11. _____

12. In a fractional expression, the sum of the values of the numerator and the denominator is 23. The product of their values is 112. Find the values of the numerator and the denominator.

12. _____

13. _____

13. A rectangle with diagonal of length $\sqrt{241}$ has an area of 60. Find the dimensions of the rectangle.

14. _____

14. Find two numbers whose product is 96 if the sum of their squares is 292.

15. A garden contains two square flower beds. Find the length of each bed if the sum of their areas is 137 ft^2 and the difference of their areas is 105 ft^2.

15. _____

NAME _____

ANSWERS	
	Classify the equation as a circle, an ellipse, a parabola, or a hyperbola.
16. _____	16. $9 - xy + 10y^2 = 10y^2$
17. _____	17. $x^2 = 15y - 9$
18. _____	18. $\dfrac{(x - 4)^2}{121} + \dfrac{(y + 1)^2}{64} = 1$
19. _____	19. $9y^2 - x^2 - 18y - 2x = 1$
20. _____	20. $y = 2x - 15 - 3x^2$
21. _____	21. $x^2 + y^2 + 6x - 2y - 90 = 0.$

22. Find an equation of a hyperbola having $y = \frac{2}{3}x$ and $y = -\frac{2}{3}x$ as asymptotes and one vertex $(0,-2)$.

22. _____

23. Find an equation of the following parabola: Line of symmetry parallel to the x-axis, vertex $(-3,2)$, and passing through $(1,6)$.

23. _____

24. The sum of two numbers is 10 and the product is 14. Find the sum of the reciprocals of the numbers.

24. _____

NAME _____

CLASS _____ SCORE _____ GRADE _____

MATCHING:
Place the letter of the answer in the answer blank. Some
letters may be used more than once, and some will not be
used.

ANSWERS

Find:

Answers:

1. An equation of the ellipse
   with vertices $(-6,-2)$,
   $(12,-2)$, $(3,1)$, $(3,-5)$.

   a) $(x - 3)^2 + (y + 2)^2 = 1$

   b) $\dfrac{y^2}{9} - \dfrac{x^2}{81} = 1$

2. One of the foci of the
   ellipse $\dfrac{x^2}{9} + \dfrac{y^2}{16} = 1$.

   c) $\dfrac{(x - 3)^2}{81} + \dfrac{(y + 2)^2}{9} = 1$

   d) $\left(0, \sqrt{7}\right)$

3. One of the vertices of the
   hyperbola $16x^2 - 9y^2 = 144$.

   e) $(0,5)$

   f) $(0,4)$

4. An equation of the
   hyperbola having asymptotes
   $y = -\dfrac{1}{3}x$ and $y = \dfrac{1}{3}x$ and one
   vertex $(0,-3)$.

   g) $(x - 3)^2 + (y + 2)^2 = 81$

   h) $81y^2 - 9x^2 = 1$

   i) $x^2 - 32y + 96 = 0$

5. The vertex of the parabola
   $x^2 - 16y + 16 = 0$.

   j) $(0,1)$

   k) $(10,1)$

6. An equation of the parabola
   with directrix $y = -5$ and
   focus $(0,11)$.

   ℓ) $\dfrac{(x - 3)^2}{81} + \dfrac{(y + 2)^2}{9} = 1$

   m) $\left(-\sqrt{7}, 0\right)$

   n) $(3,0)$

   o) $y^2 - 32x + 6x - 11 = 0$

Classify each equation as a circle, ellipse, parabola,
or hyperbola.

7. $y^2 + 4x - 8 = 0$

8. $25x^2 = 9y^2 - 225$

9. $3x^2 + 3y^2 - 6x + 18y + 3 = 0$

10. $x^2 + 4y^2 + 32y + 48 = 0$

1. _____

2. _____

3. _____

4. _____

5. _____

6. _____

7. _____

8. _____

9. _____

10. _____

NAME _____

ANSWERS	
11.	See graph.
12.	See graph.
13.	See graph.
14.	See graph.
15.	
16.	
17.	
18.	

11. Graph:

$$10x^2 + 13xy - 3y^2 = 0.$$

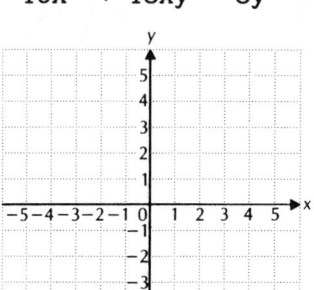

12. Graph: $xy = -6$.

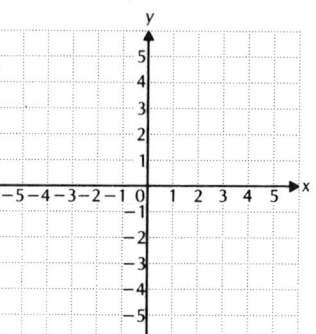

13. Graph:

$$x^2 - 4x + 8y + 4 = 0.$$

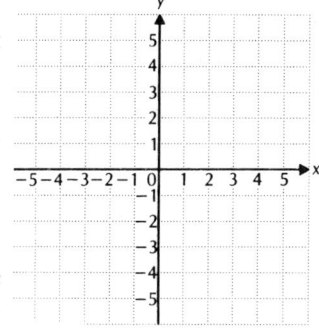

14. Graph:

$$4(x - 1)^2 - 9(y - 2)^2 = 36.$$

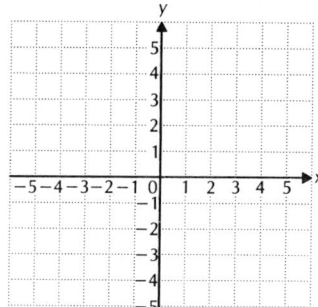

15. Solve: $x^2 + 4y^2 = 16,$
$\qquad x + 2y = 4$

16. Solve: $2x^2 - y^2 = 4,$
$\qquad x^2 + 2y^2 = 12$

17. Find two numbers whose product is -72 if the sum of their squares is 180.

18. Find an equation of the parabola containing $(-2,1)$ whose line of symmetry is parallel to the x-axis, and whose vertex is $(3,-4)$.

NAME _____

CLASS _____ SCORE _____ GRADE _____

Write the letter of your response on the answer blank.

1. Graph: $5x^2 - 8xy + 3y^2 = 0$.

1. _____

a)
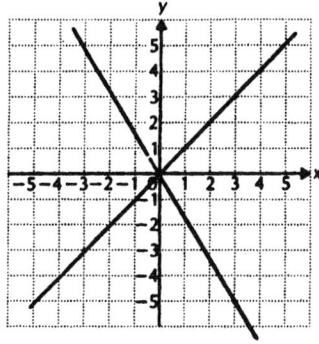

b)
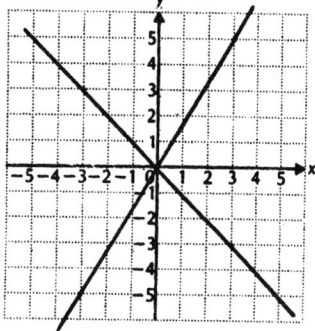

c)
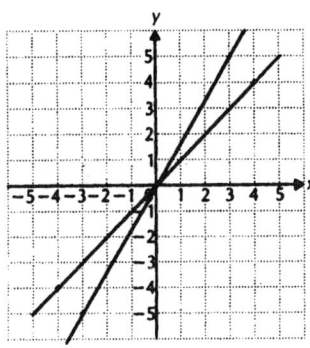

2. _____

2. Find the center, vertices, and foci of the ellipse
$49x^2 + 4y^2 - 294x + 8y + 249 = 0$.

a) C: (3,−1); V: (1,−1), (3,−8), (5,−1), (3,6);
   F: $\left(3,-1 + 3\sqrt{5}\right)$, $\left(3,-1 - 3\sqrt{5}\right)$

b) C: (3,−1); V: (4,0), (−4,0), (0,7), (0,−7);
   F: $\left(0,3\sqrt{5}\right)$, $\left(3,- 3\sqrt{5}\right)$

c) C: (−3,1); V: (0,1), (−7,1), (−3,6), (−3,−8);
   F: $\left(-2,1 - 3\sqrt{5}\right)$, $\left(-2,1 + 3\sqrt{5}\right)$

d) C: (3,−1); V: (7,0), (−7,0), (4,0), (−4,0);
   F: $\left(3\sqrt{5},0\right)$, $\left(-3\sqrt{5},0\right)$

ANSWERS

3. _____

4. _____

5. _____

3. Find an equation of the ellipse with vertices $(-6,-2)$, $(12,-2)$, $(3,8)$, and $(3,-12)$.

a) $\dfrac{x^2}{81} - \dfrac{y^2}{100} = 1$

b) $\dfrac{(x-3)^2}{81} + \dfrac{(y+2)^2}{100} = 1$

c) $\dfrac{x^2}{9} - \dfrac{y^2}{10} = 1$

d) $\dfrac{(x+3)^2}{81} - \dfrac{(y-2)^2}{100} = 1$

4. Graph: $x^2 + 4y^2 = 4$.

a)

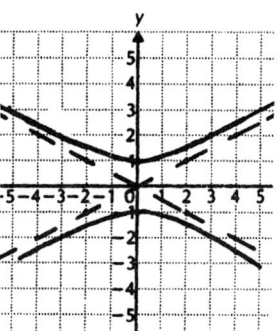

b)

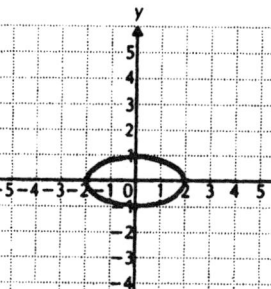

c)

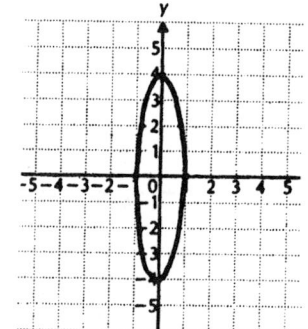

5. Find the vertices, foci, and asymptotes of the hyperbola $9y^2 - 16x^2 = 144$.

a) V: $(4,0)$, $(-4,0)$; F: $(8,0),(-8,0)$;
   A: $y = \dfrac{3}{4}x$, $y = -\dfrac{3}{4}x$

b) V: $(0,3)$, $(0,-3)$; F: $(0,9)$, $(0,-9)$;
   A: $y = 5x$, $y = -5x$

c) V: $(0,4)$, $(0,-4)$; F: $(0,8)$, $(0,-8)$;
   A: $y = \dfrac{1}{5}x$, $y = -\dfrac{1}{5}x$

d) V: $(0,4)$, $(0,-4)$; F: $(0,5)$, $(0.-5)$;
   A: $y = \dfrac{4}{3}x$, $y = -\dfrac{4}{3}x$

NAME _____

6. Graph: $xy = 6$.

a)   b)   c)

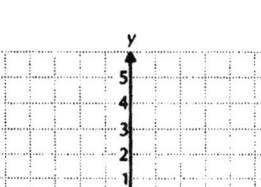
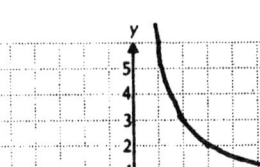
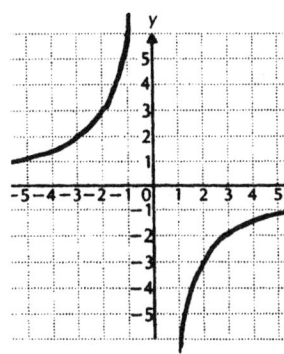

6. _____

7. _____

7. Find an equation of the parabola with directrix $x = -2$ and focus $(4,-2)$.

a) $(y + 2)^2 = 8(x + 1)$ b) $(y + 2)^2 = 12(x - 1)$

c) $y^2 = 16x$ d) $(x + 4)^2 = 4(y - 1)$

8. _____

8. Find the directrix of the parabola $x^2 - 14x - 40y - 31 = 0$.

a) $y = -1$ b) $x = -3$ c) $x = 7$ d) $y = -12$

9. Graph: $x^2 - 8y - 8 = 0$.

a)   b)   c)

9. _____

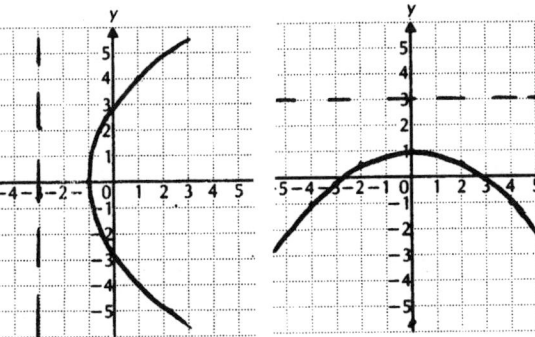

10. _____

10. Classify the equation: $x^2 - 125 = 25y^2 - 100y$.

a) circle   b) ellipse   c) hyperbola   d) parabola

NAME _____

11. _____

11. Classify the equation: $4x^2 + 2x + 1 = 5y - 4y^2$.

    a) circle   b) ellipse   c) hyperbola d) parabola

12. Solve: $x + 5y = 6$,
            $2x^2 + 3y = 5$.

The solution set

    a) is empty.

    b) consists of two ordered pairs.

12. _____

    c) consists of one ordered pair.

    d) consists of four ordered pairs.

13. Solve: $x^2 + y^2 = 4$,
            $36x^2 + 4y^2 = 144$.

The solution set

13. _____

    a) is empty.

    b) consists of two ordered pairs.

    c) consists of one ordered pair.

    d) consists of four ordered pairs.

14. The sum of two numbers is −3, and the difference of their squares is 57. Find the larger of the two numbers.

14. _____

    a) 8   b) 19   c) −11   d) 11

15. A rectangle has a perimeter of 40 cm and the length of a diagonal is $4\sqrt{13}$ cm. Find the area of the rectangle.

    a) 40 cm^2 b) 99 cm^2 c) 96 cm^2 d) 13 cm^2

15. _____

NAME _____

	ANSWERS

1. Find the 11th term of the arithmetic sequence
$\frac{1}{3}$, 1, $\frac{5}{3}$, $\frac{7}{3}$, ...

2. The 4th term of an arithmetic sequence is 17, and the 10th term is 41. Find the common difference.

3. Which term of the arithmetic sequence $\frac{1}{4}$, 1, $\frac{7}{4}$, $\frac{5}{2}$, . . . is 16?

4. Insert three arithmetic means between 5 and 12.

5. Find the 10th term of the geometric sequence
$\frac{1}{2}$, $-\frac{1}{4}$, $\frac{1}{8}$, $-\frac{1}{16}$, ...

6. Evaluate the sum $\sum_{k=1}^{6} 4\left(\frac{1}{2}\right)^{k-1}$.

7. Which of the following infinite geometric sequences have sums?

a) 4, −20, 100, −500, . . .

b) 2, 0.8, 0.32, 0.128, . . .

c) $-\frac{1}{3}$, $\frac{2}{3}$, $-\frac{4}{3}$, 2, . . .

8. Find the sum of the infinite geometric sequence
36, −6, 1, $-\frac{1}{6}$, ...

9. If John saves $15 one week, $25 the second week, $35 the third weeek, and continues this pattern, how much money would he save in the 30th week? What is the sum of his savings?

10. A rubber ball is dropped from a height of 27 ft and always rebounds $\frac{1}{3}$ of the distance of the previous fall. What distance does it rebound on the 7th time?

11. Find fractional notation for $0.177\overline{7}$.

12. Use mathematical induction. Prove that for every natural number n,

$1 \cdot 2 + 2 \cdot 4 + 3 \cdot 8 + \ldots + n \cdot 2^n = (n - 1)2^{n+1} + 2.$

ANSWERS

1. _____

2. _____

3. _____

4. _____

5. _____

6. _____

7. _____

8. _____

9. _____

10. _____

11. _____

12. See work.

NAME _____

ANSWERS

13. _____

14. a) _____

  b) _____

  c) _____

15. _____

16. _____

17. _____

18. _____

19. _____

20. _____

21. _____

22. _____

23. _____

24. _____

25. _____

13. Find the first 4 terms of this recursively defined sequence.
$$a_1 = 3, \quad a_{k+1} = 1 + a_k^2$$

14. How many code symbols can be formed using 4 out of 7 of the letters A, B, C, D, E, F, G if the letters:

  a) can be repeated?

  b) cannot be repeated?

  c) cannot be repeated but must begin with B?

15. A 5-member team is to be chosen from 11 basketball players. In how many ways can this be done?

16. How many committees can be formed from a group of 8 governors and 9 senators if each committee contains 5 governors and 6 senators?

17. In how many distinguishable ways can the letters of the word ALGEBRA be arranged?

18. Determine the number of subsets of a set of 7 numbers.

19. Find the 4th term of $(3a - b)^6$.

20. Expand $\left(x + \sqrt{3}\right)^4$.

21. What is the probability of getting a total of 5 on a roll of a pair of dice?

22. From a deck of 52 cards, 1 card is drawn. What is the probability of drawing a ten or a queen?

23. From a bag containing 6 nickels, 9 dimes, and 5 quarters, 8 coins are drawn at random, all at once. What is the probability of getting 3 nickels, 4 dimes and 1 quarter?

24. Find decimal notation, rounded to six decimal places, for the first 5 terms of $a_n = \left(\dfrac{2n + 1}{n}\right)^n$.

25. How many games are played in a league with 10 teams if each team plays each other team once?

CLASS _____ SCORE _____ GRADE _____

---

1. Find the 18th term of the arithmetic sequence
$$\frac{2}{5}, \frac{6}{5}, 2, \frac{14}{5}, \ldots$$

ANSWERS

1. _____

2. The 2nd term of an arithmetic sequence is 9, and the 8th term is 51. Find the common difference.

2. _____

3. Which term of the arithmetic sequence
$$\frac{2}{3}, 2, \frac{10}{3}, \frac{14}{3}, \ldots \text{ is } \frac{38}{3}?$$

3. _____

4. Insert three arithmetic means between 6 and 21.

4. _____

5. Find the 6th term of the geometric sequence
0.3, 0.6, 1.2, 2.4, ...

6. Evaluate the sum $\displaystyle\sum_{k=1}^{5} \left(\frac{2}{3}\right)^{k-1}$.

5. _____

7. Which of the following infinite geometric sequences have sums?

   a) $\frac{1}{2}, \frac{3}{2}, \frac{9}{2}, \frac{27}{2}, \ldots$

   b) 0.24, −0.12, 0.06, −0.03, . . .

   c) 46, 4.6, 0.46, 0.046, . . .

6. _____

7. _____

8. Find the sum of the infinite geometric sequence
$$1, \frac{1}{3}, \frac{1}{9}, \frac{1}{27}, \ldots$$

8. _____

9. A bomb drops from a plane and falls 16 feet the first second, 48 feet the second second, 80 feet the third second, and so on. Find the total number of feet the bomb fell in 18 seconds.

9. _____

10. A college student borrows $1000 at 10% interest compounded annually. The loan is paid off at the end of 4 years. How much is paid back?

10. _____

11. Find fractional notation for $0.2424\overline{24}$.

11. _____

12. _____ 12. See work.

12. Use mathematical induction. Prove that for every natural number n,
$$\frac{1}{1\cdot2\cdot3} + \frac{1}{2\cdot3\cdot4} + \cdots + \frac{1}{n(n+1)(n+2)} = \frac{n(n+3)}{4(n+1)(n+2)}.$$

TEST FORM B

ANSWERS	
13. _____	13. Find the first 4 terms of this recursively defined sequence. $a_1 = -4$, $a_{k+1} = 9 - 2a_k$
14. a) _____ b) _____ c) _____	14. How many code symbols can be formed using 3 out of 8 of the letters B, D, M, P, Q, X, Y, Z if the letters: a) can be repeated? b) cannot be repeated? c) cannot be repeated but must end with XY?
15. _____	15. From a group of 12 gymnasts, 4 are to be chosen to represent their school. In how many ways can the representative be chosen?
16. _____ 17. _____	16. Of the first 8 questions on a test a student must answer 5. Of the next 6 questions the student must answer 3. In how many ways can this be done?
18. _____	17. In how many distinguishable ways can the letters of the word CALCULUS be arranged?
19. _____	18. Determine the number of subsets of a set of 9 numbers.
20. _____	19. Find the 3rd term of $(4a - b)^7$. 20. Expand $\left(\sqrt{5} - a\right)^5$.
21. _____	21. What is the probability of getting a total of 9 on a roll of a pair of dice?
22. _____	22. From a deck of 52 cards, 1 card is drawn. What is the probability of drawing a diamond?
23. _____	23. From a group of 9 men and 7 women, a committee of 6 is chosen. What is the probability that 3 men and 3 women will be chosen?
24. _____	24. The value of an office machine is $4000. Its scrap value is 65% of its value the year before. Give a sequence that lists the scrap value of the machine for each year of a 5-year period.
25. _____	25. Solve $_nP_8 = 9 \cdot \, _{n-1}P_7$ for n.

TEST FORM C   CLASS _____ SCORE _____ GRADE _____

	ANSWERS

1. Find the 10th term of the arithmetic sequence
   $\frac{5}{8}$, 1, $\frac{11}{8}$, $\frac{7}{4}$, ...

   1. _____

2. The 3rd term of an arithmetic sequence is 19, and the 9th term is 67. Find the common difference.

   2. _____

3. Which term of the arithmetic sequence
   $\frac{3}{8}$, $\frac{7}{8}$, $\frac{11}{8}$, $\frac{15}{8}$, . . . is $\frac{51}{8}$?

   3. _____

4. Insert three arithmetic means between 4 and 15.

   4. _____

5. Find the 7th term of the geometric sequence
   6, -3, $\frac{3}{2}$, $-\frac{3}{4}$

   5. _____

6. Evaluate the sum $\sum\limits_{k=1}^{5} (3 - 2^k)$.

   6. _____

7. Which of the following infinite geometric sequences have sums?

   a) 8, -2, $\frac{1}{2}$, $-\frac{1}{8}$, . . .

   b) 21, 52.5, 131.25, 328.125, . . .

   c) $\frac{7}{8}$, $\frac{7}{4}$, $\frac{7}{2}$, 7, . . .

   7. _____

8. Find the sum of the infinite geometric sequence
   7, 0.7, 0.07, 0.007, . . .

   8. _____

9. A hiker walks 3.9 mi the first day, 4.7 mi the second day, 5.5 mi the third day, and so on. What is the total distance hiked in 2 weeks?

   9. _____

10. Grantville has a population of 20,000 now and the population is increasing 8% every year. What will the population be in 5 years?

    10. _____

11. Find fractional notation for 8.5555.

    11. _____

12. Use mathematical induction. Prove that for every natural number n,
    $$\frac{1}{3} + \frac{1}{15} + \frac{1}{35} + . . . + \frac{1}{4n^2 - 1} = \frac{n}{2n + 1}.$$

    12. __See work.__

NAME _____

ANSWERS	
13. _____	13. Find the first 4 terms of this recursively defined sequence. $a_1 = 4$, $a_{k+1} = 2a_k - 3$
14. a) _____ b) _____ c) _____	14. How many code symbols can be formed using 5 out of 6 of the letters C, D, E, F, G, H if the letters: a) can be repeated? b) cannot be repeated? c) cannot be repeated but must begin with EFG?
15. _____	15. There are 20 students in a fraternity. How many sets of 4 officers may be selected?
16. _____	16. A committee is to be formed from a group of 12 women and 10 men and is to consist of 5 women and 5 men. How many committees can be formed?
17. _____	17. In how many distinguishable ways can the letters of the word SCIENCE be arranged?
18. _____	18. Determine the number of subsets of a set of 5 numbers.
19. _____	19. Find the 6th term of $(a - 3b)^6$.
20. _____	20. Expand $\left(\sqrt{7} + a\right)^6$.
21. _____	21. What is the probability of getting a total of 7 on a roll of a pair of dice?
22. _____	22. From a deck of 52 cards, 1 card is drawn. What is the probability of drawing a 5 or an ace?
23. _____	23. If 7 marbles are drawn at random, all at once, from a bag containing 6 green marbles, 8 red marbles, and 3 white marbles, what is the probability that 3 will be green and 4 will be red?
24. _____	24. Find four numbers in an arithmetic sequence such that the first plus twice the second is 12, and the fourth minus the third is 3.
25. _____	25. Solve for n: $\begin{pmatrix} n \\ n - 2 \end{pmatrix} = 45$.

NAME _____

CLASS _____ SCORE _____ GRADE _____

	ANSWERS

1. Find the 16th term of the arithmetic sequence
   $\frac{3}{4}$, $\frac{3}{2}$, $\frac{9}{4}$, 3, ...

1. _____

2. The 3rd term of an arithmetic sequence is 16, and the 8th term is 41. Find the common difference.

2. _____

3. Which term of the arithmetic sequence 4, $\frac{11}{2}$, 7, $\frac{17}{2}$, . . is 43?

3. _____

4. Insert three arithmetic means between 3 and 18.

4. _____

5. Find the 9th term of the geometric sequence
   36, 12, 4, $\frac{4}{3}$, ...

5. _____

6. Evaluate the sum $\sum\limits_{k=1}^{4} 10\left(\frac{1}{5}\right)^k$.

6. _____

7. Which of the following infinite geometric sequences have sums?
   a) $\frac{1}{4}$, $\frac{1}{8}$, $\frac{1}{16}$, $\frac{1}{32}$, . . .
   b) 1, 1.2, 1.44, 1.728, . . .
   c) -8, -1, $-\frac{1}{8}$, $-\frac{1}{64}$, . . .

7. _____

8. Find the sum of the infinite geometric sequence
   -8, 2, $-\frac{1}{2}$, $\frac{1}{8}$, ...

8. _____

9. A theater has 35 seats in the first row, 50 seats in the second row, 65 seats in the third row, and so on, for 20 rows. How many seats are there in the theater?

9. _____

10. _____

10. Suppose someone offered you a job for 2 weeks under the following conditions. You will be paid $1.00 for the first day, $2.00 for the second day, $4.00 for the third day, and so on. How much would you earn altogether?

11. _____

11. Find fractional notation for $5.45\overline{45}$.

12. See work. _____

12. Use mathematical induction. Prove that for every natural number n,
    $1^3 + 3^3 + 5^3 + . . . + (2n - 1)^3 = n^2(2n^2 - 1)$.

NAME _____

ANSWERS	
13. _____	13. Find the first 4 terms of this recursively defined sequence. $$a_1 = 4, \quad a_{k+1} = 5a_k - 2$$
14. a) _____	14. How many code symbols can be formed using 4 out of 9 of the letters A, B, C, D, F, G, M, Y, Z if the letters:
b) _____	a) can be repeated?
c) _____	b) cannot be repeated?
	c) cannot be repeated but must end with MY?
15. _____	15. On a test, a student is to select 7 out of 12 questions. In how many ways can she do this?
16. _____	16. A restaurant list 8 entrees and 6 vegetables on its menu. If you choose 1 entree and 3 vegetables, how many ways can you make this choice?
17. _____	17. In how many distinguishable ways can the letters of the word ZOOLOGY be arranged?
18. _____	18. Determine the number of subsets of a set of 6 numbers.
19. _____	19. Find the 3rd term of $(2x - 3y)^5$.
20. _____	20. Expand $\left(x - \sqrt{6}\right)^5$.
21. _____	21. What is the probability of getting a total of 10 on a roll of a pair of dice?
22. _____	22. From a deck of 52 cards, 1 card is drawn. What is the probability of drawing a red card?
23. _____	23. From a group of 7 men and 10 women, a committee of 5 is chosen. What is the probability that 2 men and 3 women will be chosen?
24. _____	24. Find three numbers in an arithmetic sequence such that the sum of the first and third is 8 and the product of the first and second is 24.
25. _____	25. Solve $_nP_4 = 5 \cdot {}_{n-1}P_3$ for n.

NAME _____

CLASS _____ *SCORE* _____ *GRADE* _____

ANSWERS

1. Find the 32nd term of the arithmetic sequence
   6, 15, 24, ...

   1. _____

2. The 3rd term of an arithmetic sequence is 9, and the
   9th term is 33. Find the 15th term.

   2. _____

3. Which term of the arithmetic sequence $\frac{7}{2}$, 4, $\frac{9}{2}$, . . .
   is 28?

   3. _____

4. Find the sum of the first 10 terms of the arithmetic
   sequence q - 3p, q - 2p, q - p, ...

   4. _____

5. Find the 7th term of the geometric sequence
   0.2, 0.6, 1.8, ...

   5. _____

6. Find the sum of the first 6 terms of the geometric
   sequence $\frac{1}{8}$, $-\frac{1}{12}$, $\frac{1}{18}$, ...

   6. _____

7. Evaluate the sum $\sum_{k=1}^{5} \frac{4}{2^{k-1}}$.

   7. _____

8. Insert five arithmetic means between 6 and 22.

   8. _____

9. If you save $5 the first week, $7 the second week,
   $9 the third week, and so on, how much do you save in
   52 weeks?

   9. _____

10. A ball dropped from the top of a tower (400 m high)
    always rebounds 25% of the distance of the previous
    fall. How far has the ball traveled when it hits the
    ground for the third time?

    10. _____

    11. _____

11. A city has a population of 350,000 now, and the
    population is only increasing 2% every year. What will
    the population be in 5 years?

    12. _____

12. Find fractional notation for $4.02\overline{2}$.

TEST FORM E

ANSWERS	

13. See work. _____

14. _____

15. _____

16. _____

17. a) _____

b) _____

c) _____

18. _____

19. _____

20. _____

21. _____

22. _____

23. _____

13. Use mathematical induction. Prove for every natural number n.
$$1 + 3 + 6 + \ldots + \frac{1}{2}n(n + 1) = \frac{n(n + 1)(n + 2)}{6}$$

14. The winner in a bike race gets to select 1 prize from Group A, 2 prizes from Group B, and 1 prize from Group C. If Group A contains 4 trips, Group B contains 6 bicycles, and Group C contains 8 electronic games, how many different selections can be made?

15. In how many ways can 7 students be arranged in a straight line? in a circle?

16. Suppose the expression $x^4y^3z^2$ is expressed without exponents. In how many ways can this be done?

17. How many 4-digit numbers can be named using the digits 1, 2, 3, 4 and 5 if the digits:

a) can be repeated?

b) are not repeated?

c) are not repeated and the numbers must be less than 4000?

18. If there are 8 outside doors in a school, in how many ways can a student enter one door and leave by a different door?

19. Find the 5th term of $\left(5x - \sqrt{2y}\right)^7$.

20. Expand $(2x + y^2)^5$.

21. What is the probability of getting a total of 11 on a roll of a pair of dice?

22. From a deck of 52 cards, 2 cards are drawn without replacement. What is the probability that both are 5's?

23. From a group of 12 senators and 15 representatives, a committee of 5 is chosen. What is the probability that 2 senators and 3 representatives will be chosen?

NAME _____

CLASS _____ SCORE _____ GRADE _____

Write the letter of your response on the answer blank.   ANSWERS

1. Which term of the arithmetic sequence   1. _____
   1, $\frac{5}{2}$, 4, $\frac{11}{2}$, . . . is 37?

   a) 25th   b) 42nd   c) 15th   d) 26th

2. The 5th term of an arithmetic sequence is 17 and the   2. _____
   12th term is 66. Find the common difference.

   a) 3   b) $\frac{21}{6}$ c) 7   d) 21

3. Find the sum of the first 18 terms of the arithmetic   3. _____
   sequence 2, 7, 12, . . .

   a) 783   b) 87   c) 1692   d) 801

4. Insert three arithmetic means between 10 and 75.

   a) 23, 36, 49   b) $\frac{105}{4}$, $\frac{85}{2}$, $\frac{235}{4}$ 4. _____

   c) 30, 45, 60   d) 25, $\frac{63}{2}$, 60

5. Find the 6th term of the geometric sequence   5. _____
   16, −4, 1, ...

   a) $-\frac{1}{64}$ b) $\frac{1}{4}$ c) $-\frac{3}{8}$ d) $-\frac{3}{256}$

6. Find the sum of the first 6 terms of the geometric
   sequence $\frac{1}{3}$, $-\frac{2}{3}$, $\frac{4}{3}$, $-\frac{8}{3}$, ...   6. _____

   a) $\frac{11}{3}$ b) −7   c) $\frac{20}{3}$ d) −28

7. Find the sum of the infinite geometric sequence
   3, 1, $\frac{1}{3}$, ...   7. _____

   a) 1.5   b) 1   c) 4.5   d) 2

ANSWERS	
8. _____	

8. Find $\displaystyle\sum_{k=1}^{5} 3k^{k-1}$.

   a) 701   b) 3512   c) 1875   d) 2103

9. _____

9. Find fractional notation for $0.823\overline{23}$.
   Simplify the fraction. The sum of the numerator and
   the denominator is

   a) 122   b) 343   c) 361   d) 1822

10. _____

10. An employee of a company makes deposits in the Credit
    Union as follows: $15.50 the first month, $17.00 the
    second month, $18.50 the third month, and so on, for
    three years. Find the sum of the deposits.

    a) $1503   b) $1224   c) $2322   d) $1937.50

11. _____

11. A ball dropped from the top of a tower (520 ft high)
    always rebounds $\frac{3}{5}$ of the distance of the previous fall.
    How high will the fourth bounce be?

    a) 187.2 ft   b) 67.392 ft   c) 52 ft   d) 112.32 ft

12. See work.

12. Use mathematical induction. Prove for every natural
    number n.
    $$2 + 6 + 18 + \ldots + 2(3^{n-1}) = 3^n - 1$$

13. _____

13. In how many ways can 6 floats be lined up for a parade?

    a) 2^6 b) 1   c) 6!   d) 6

	ANSWERS

14. How many different committees consisting of 2 men and 1 woman can be formed from a group of 6 men and 4 women?

    a) 2024   b) 120   c) $\dfrac{3}{10}$ d) 60

14. _____

15. How many ways can 9 people sit at a round table?

    a) ${}_9C_8$ b) ${}_9P_7$ c) $8!$ d) $9!$

15. _____

16. How many 3-digit numbers can be named using the digits 0, 2, 4, 6, 8 if the digits are not repeated and the numbers must be greater than 500?

    a) 10   b) 24   c) 12   d) 60

16. _____

17. In how many distinquishable ways can the letters of the word TALLAHASSEE be arranged?

    a) $\dbinom{11}{6}$ b) $\dfrac{6!}{3!2!2!2!}$ c) ${}_{11}P_6$ d) $\dfrac{11!}{3!2!2!2!}$

17. _____

18. How many baseball games can be played in a 14-team league if each team plays all other teams once?

    a) 91   b) 28   c) 182   d) 46

19. Expand $\left(y - \dfrac{2}{y}\right)^5$.

    a) $y^2 - 5y^3 + 10y - \dfrac{10}{y} + \dfrac{5}{y^3} - \dfrac{1}{y^5}$

18. _____

    b) $y^2 + 5y^3 + 10y + \dfrac{10}{y} + \dfrac{5}{y^3} + \dfrac{1}{y^5}$

    c) $y^5 - 10y^3 + 40y - \dfrac{80}{y} + \dfrac{80}{y^3} - \dfrac{32}{y^5}$

    d) $y^5 + 10y^3 + 40y + \dfrac{80}{y} + \dfrac{80}{y^3} + \dfrac{32}{y^5}$

19. _____

NAME _____

ANSWERS	

20. _____

21. _____

22. _____

23. _____

24. _____

25. _____

20. Find the 4th term of $(3x - 4y)^8$.

   a) $120,960x^5y^4$ b) $-870,912x^5y^3$

   c) $96,768x^4y^4$ d) $15,120x^4y^4$

21. Determine the number of subsets of a set of 9 members.

   a) $_9P_2$ b) 9^2 c) $\binom{9}{2}$ d) 2^9

22. From a deck of 52 cards, 1 card is drawn. What is the probability that it is an ace or a four?

   a) $\dfrac{8}{13}$ b) $\dfrac{2}{13}$ c) $\dfrac{1}{26}$ d) $\dfrac{1}{13}$

23. What is the probability of rolling a 4 on a roll of a pair of dice?

   a) 0   b) $\dfrac{1}{9}$ c) 6   d) $\dfrac{1}{12}$

24. If 3 marbles are drawn at random, all at once, from a bag containing 5 white marbles and 6 red marbles, what is the probability that 1 will be white and 2 will be red?

   a) $\dfrac{1}{2}$ b) $\dfrac{3}{11}$ c) $\dfrac{5}{11}$ d) $\dfrac{3}{5}$

25. Suppose 4 cards are drawn without replacement from a well-shuffled deck of 52 cards. What is the probability that all four are spades?

   a) $\dfrac{_4P_4}{_{52}P_4}$ b) $\dfrac{1}{13}$ c) $\dfrac{1}{4}$ d) $\dfrac{\binom{13}{4}}{\binom{52}{4}}$

Consider the numbers:

$15, -\dfrac{3}{2}, -\sqrt[5]{8}, 0.73, 0, \sqrt{40}, -3.7, 6.\overline{12}, -18.$

1. Which are whole numbers?

2. Which are irrational numbers?

3. Which are natural numbers?

4. Which are integers?

Compute. Write scientific notation for each answer.

5. $(4.3 \times 10^5)(3.1 \times 10^{-11})$

6. $\dfrac{4.8 \times 10^{-15}}{1.5 \times 10^{-10}}$

7. Add and simplify: $\dfrac{x}{x^2 + x - 6} + \dfrac{4}{x^2 - 4}.$

Solve.

8. $3x - 11 \geq 8x + 2$

9. $\dfrac{3}{x^2} - \dfrac{1}{x} = 14$

10. Write a quadratic equation whose solutions are $\sqrt{3}$ and $-2\sqrt{3}$.

11. Find an equation of variation in which y varies jointly as x and z and inversely as the square of w, and $y = 225$ when $x = 3$, $z = 2$, and $w = 8$.

12. Determine the meaningful replacements in the radical expression $\sqrt{18 - 2x}.$

13. Find an equation of the line through $(-2,3)$ with $m = -4$.

14. Find the distance between $(3,-2)$ and $(5,1)$.

ANSWERS

1. _____

2. _____

3. _____

4. _____

5. _____

6. _____

7. _____

8. _____

9. _____

10. _____

11. _____

12. _____

13. _____

14. _____

ANSWERS

15. _____

_____

_____

_____

_____

_____

16. _____

_____

_____

_____

_____

_____

_____

_____

17. _____

18. _____

19. _____

20. _____

21. _____

22. _____

Consider the functions f and g given by $f(x) = 2x - 4$ and $g(x) = x^2 + 3$ for Questions 15 and 16.

15. Find $(f + g)(x)$, $(f - g)(x)$, $fg(x)$, $(f/g)(x)$, $f \circ g(x)$, and $g \circ f(x)$.

16. Find the domain of f, g, f + g, f - g, fg, f/g, f∘g, and g∘f.

Use the following for Questions 17 and 18.

a) 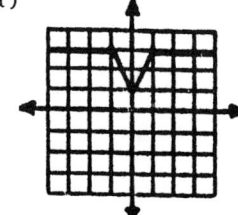   b) 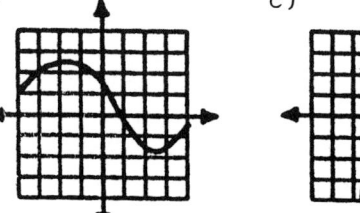   c)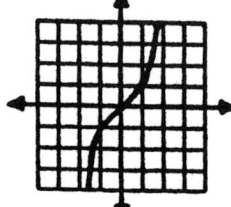

17. Which are increasing?

18. Which are neither increasing nor decreasing?

Solve.

19. $|2x - 5| \leq 7$ 20. $6x^2 + 19x - 7 < 0$

21. The sum of the base and height of a triangle is 30 cm. Find the dimensions for which the area is a maximum.

22. Factor the polynomial P(x). Then solve the equation $P(x) = 0$.

$$P(x) = x^3 + x^2 - 7x - 3$$

23. Find a polynomial of lowest degree having roots 2 and −3, and having −1 as a root of multiplicity 2 and 4 as a root of multiplicity 3.

24. Graph: $f(x) = \dfrac{x^2 - 5x + 5}{x - 2}$.

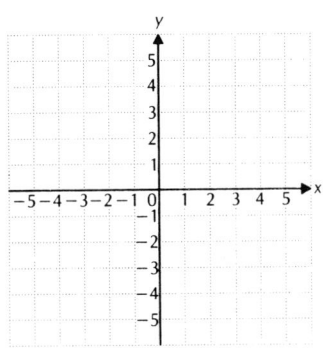

25. Write an equation of the inverse of the relation $y = 2x - 5$.

26. Find $\log_4 13$ using natural logarithms.

27. Solve: $\log_{\sqrt{8}} x = 4$.

28. How many years will it take an investment of $6000 to double if interest is compounded annually at 8%?

Find using a calculator.

29. log 15,750   30. ln 2.35

ANSWERS

23. _____

24. _See graph._

25. _____

26. _____

27. _____

28. _____

29. _____

30. _____

NAME _____

ANSWERS	
31. _____	31. Solve: $0.9x - 1.3y = -0.3,$ $0.06x - 0.07y = 0.03.$
32. _____	32. Solve using matrices: $x - z = 1,$ $3x + y + z = 8,$ $x - y - z = 0.$
33. _____	33. Find numbers a, b, and c such that the function $f(x) = ax^2 + bx + c$ fits the data points $(0,-4)$, $(-1,-9)$, and $(2,-6)$. Then write the equation for the function.
34. _____	34. Evaluate: $\begin{vmatrix} 5 & -1 & 3 \\ 0 & 2 & 4 \\ 1 & -1 & 5 \end{vmatrix}$.
35. _____	35. Add: $\begin{bmatrix} 5 & -1 \\ 3 & 7 \end{bmatrix} + \begin{bmatrix} 2 & 0 \\ 1 & 5 \end{bmatrix}$.
36. _____	36. Find A^{-1}, if it exists, for $A = \begin{bmatrix} 2 & 3 \\ -1 & 6 \end{bmatrix}$.
37. _____	37. Decompose into partial fractions: $\dfrac{3x + 5}{(x + 1)(x + 2)}.$

	ANSWERS

38. Find an equation of the ellipse with vertices $(-2,0)$, $(2,0)$, $(0,-3)$, $(0,3)$.

38. _____

39. Graph: $xy = -4$.

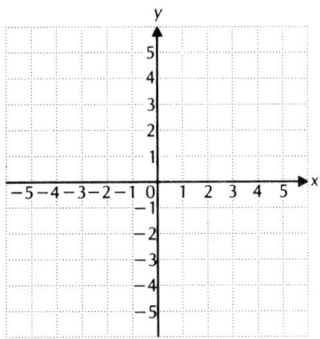

39. _See graph._

40. _____

40. The difference of two numbers is 8, and the difference of their squares is 112. What are the numbers?

41. _____

41. A rectangle has an area of 60 m^2 and a perimeter of 38 m. Find the dimensions.

42. _____

42. Classify $x^2 + y^2 - 4x + 2y = -2$ as a circle, an ellipse, a parabola, or a hyperbola.

43. A theater has 40 seats in the first row, 65 seats in the second row, 90 seats in the third row, and so on, for 30 rows. How many seats are in the theater?

43. _____

ANSWERS

44. _____

44. Evaluate the sum $\sum\limits_{k=1}^{4} 3(5 + 2^k)$.

45. a) _____

b) _____

c) _____

45. How many code symbols can be formed using 5 out of 9 of the letters A, B, C, D, E, F, G, H, I if the letters:

a) can be repeated?

b) cannot be repeated?

c) can be repeated but must begin with B?

46. _____

46. Find the first 5 terms of this recursively defined sequence.

$$a_1 = 3$$
$$a_{k+1} = 4a_k - 5$$

47. _____

47. Expand $\left(x + \sqrt{2}\right)^5$.

48. _____

48. If 5 marbles are drawn at random, all at once, from a bag containing 8 green marbles, 4 red marbles, and 5 white marbles, what is the probability that 3 will be green and 2 will be white?

NAME _____

CLASS _____ SCORE _____ GRADE _____

	ANSWERS

Compute.

1. $-5.8 - 3.6$ 2. $35 + |-35|$ 1. _____

3. $(-17)(-4)$ 4. $\dfrac{56}{-2}$ 2. _____

3. _____

Factor.

5. $6x^2 - 7x - 20$ 6. $a^3 + 343$

4. _____

7. $6x + 12y - x^2 - 2xy$

5. _____

8. Find the reciprocal of 2 + 3i and express it in the
    form a + bi.   6. _____

Solve.   7. _____

9. $12x^2 - x - 63 = 0$

8. _____

10. $\sqrt{y + 10} = \sqrt{5y - 11} - 3$

9. _____

11. The length of a rectangle is three times the width.
     The perimeter is 48 in. Find the dimensions.   10. _____

12. Determine the nature of the solutions of   11. _____
     $x^2 - 6x + 5 = 0$ by evaluating the discriminant.

12. _____

Consider the relation {(1,4), (-2,3), (7,4), (1,8)}.

13. Determine whether the relation is a function.   13. _____

14. Find the range.   14. _____

15. Find the domain.   15. _____

ANSWERS	

16. _____

17.   See graph. _____

18. _____

19.   See graph. _____

20. _____

21.   See graph. _____

22. _____

23. _____

16. Find an equation of a circle with center (5,−1) and radius 3.

17. Graph $y = \frac{2}{3}x - 4$ using the slope and the y-intercept.

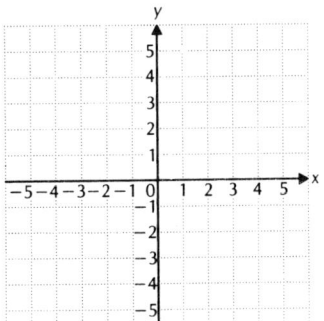

18. Write interval notation for $\{x \mid x \leq -5\}$.

19. Graph: $f(x) = x^3 - x^2 - 9x + 9$.

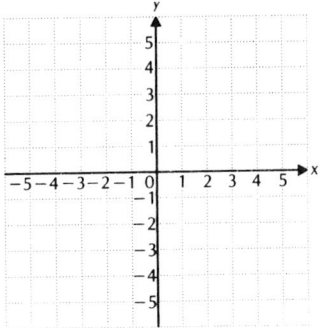

20. Find the x-intercepts of $f(x) = 2x^2 - x - 21$.

21. Graph: $\{x \mid x < -4\} \cup \{x \mid x > -2\}$.

$\longleftarrow\!\!\mid\!\mid\!\mid\!\mid\!\mid\!\mid\!\mid\!\mid\!\mid\!\mid\!\mid\!\mid\!\mid\!\mid\!\longrightarrow$

22. Determine whether x − 5 is a factor of
$x^3 - 4x^2 - 7x + 10$.

23. Find $\{3,5,7,9,13\} \cup \{1,2,5,9,15\}$.

	ANSWERS

24. Solve: $|2x + 3| = 11$.

24. _____

25. The population of a certain city was 200,000 in 1990. The exponential growth rate was 3.4% per year.

   a) Find the exponential growth function.

   b) Predict the population of the city in the year 2000.

   c) When will the population be 400,000?

25. a) _____

     b) _____

     c) _____

26. Graph: $y = \log_3 x$.

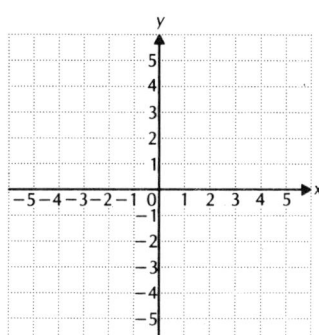

26.   See graph. _____

27. _____

27. Write an equivalent expression containing a single logarithm: $4 \log_a x - \frac{1}{2} \log_a y - 2 \log_a z$.

28. _____

29. _____

Find using a calculator.

28. $\ln 0.00027$ 29. $\log (-15.21)$

30. _____

30. The pH of a substance is 5.4. What is the hydrogen ion concentration?

NAME _____

ANSWERS

31. _____

31. A boat travels 40 km downstream in 2 hr. It travels 50 km upstream in 3 hr. Find the speed of the stream.

32. Solve using Cramer's rule: $4x - y = 3$, $2x + 3y = 7$.

32. _____

33. Graph: $2x + 4y \leq 8$.

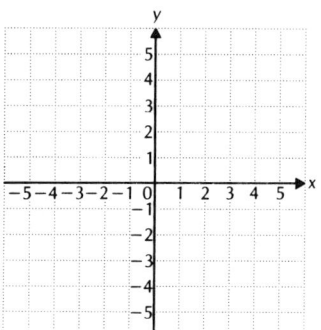

33. See graph.

34. _____

34. Let $A = \begin{bmatrix} 5 & 0 & 4 \\ 1 & -1 & 0 \\ 2 & 3 & 7 \end{bmatrix}$. Find a_{21}, M_{21}, and A_{21}.

35. Evaluate: $\begin{vmatrix} 0 & 1 & 4 \\ 3 & 2 & 5 \\ 1 & 0 & 6 \end{vmatrix}$.

35. _____

36. Given $A = \begin{bmatrix} 5 & -2 & 3 \\ 1 & 0 & 2 \end{bmatrix}$ and $B = \begin{bmatrix} 5 & 0 \\ -1 & 2 \\ 2 & 3 \end{bmatrix}$,

find AB, if possible.

36. _____

37. Classify as consistent or inconsistent, dependent or independent.

$5x + 3y = 4$,
$-5x - 3y = 0$.

37. _____

38. Graph: $2x^2 + 2xy - 12y^2 = 0$.

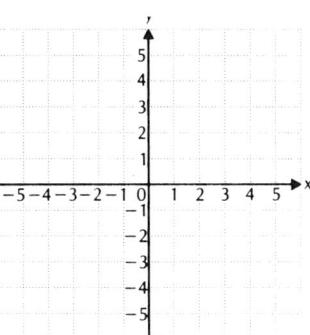

39. Find the center, vertices, foci, and asymptotes of the hyperbola $16x^2 - 9y^2 = 144$.

40. Solve: $y^2 = x + 3$,
    $5y = x + 9$.

41. Find two numbers whose product is 132 if the sum of their squares is 265.

42. Classify $x^2 - 4y^2 + 2x + 16y = 19$ as a circle, an ellipse, a parabola, or a hyperbola.

43. Which term of the arithmetic sequence
    $5, \frac{15}{2}, 10, \frac{25}{2}, \ldots$ is 35?

ANSWERS

38. See graph.

39. _____

40. _____

41. _____

42. _____

43. _____

FINAL EXAMINATION   *NAME*

TEST FORM B

ANSWERS	44. Which of the following infinite geometric sequences have sums?

44. _____

44. Which of the following infinite geometric sequences have sums?

a) $\dfrac{1}{3}, \dfrac{1}{6}, \dfrac{1}{12}, \dfrac{1}{24}, \ldots$

b) 3, 3.3, 3.63, 3.993, . . .

c) $-12, -4, -\dfrac{4}{3}, -\dfrac{4}{9}, \ldots$

45. _____

45. What is the probability of getting a total of 7 on a roll of a pair of dice?

46. _____

46. Determine the number of subsets of a set of 8 members.

47. _____

47. Find the 5th term of $(3x + 5y)^6$.

48. _____

48. From a deck of 52 cards, 1 card is drawn. What is the probability of drawing an ace or a three?

182

NAME _____

CLASS _____ SCORE _____ GRADE _____

Convert to decimal notation.

ANSWERS

1. 6.129×10^{-5} 2. 4.1×10^4

1. _____

3. 2.15 E 3   4. 1.29 E -2

2. _____

5. Divide and simplify: $\dfrac{x^2 - 4x - 5}{x^2 - x - 6} \div \dfrac{(x + 1)^2}{x^2 - 9}$.

3. _____

6. Rationalize the denominator: $\dfrac{7 + \sqrt{x}}{7 - \sqrt{x}}$.

4. _____

7. The diagonal of a square has length $11\sqrt{2}$. Find the length of a side of the square.

5. _____

Simplify.

6. _____

8. i^{15} 9. $(2 + 6i) - (-1 - 5i)$

10. $\dfrac{3 + 2i}{1 - 4i}$ 11. $(7 - i)(7 + i)$

7. _____

12. The speed of a boat in still water is 12 mph. It travels 18 mi upstream and 18 mi downstream in a total time of 4 hr. What is the speed of the current?

8. _____

9. _____

13. Which of the following are graphs of functions?

a)   b)

10. _____

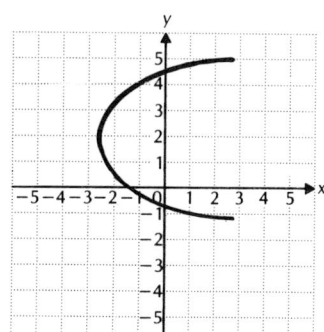   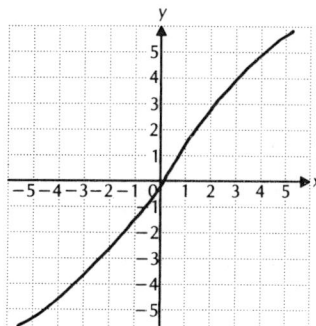

11. _____

12. _____

13. _____

ANSWERS	
14. _____	Consider the following relations for Questions 14 and 15. a) $x = y^4 - 3$ b) $x^3 - x = y^2$ c) $y^2 - 3 = x$ d) $2x - \dfrac{3}{y} = 0$ e) $3x = |y|$ f) $x = -3$
15. _____	14. Which are symmetric with respect to the origin? 15. Which are symmetric with respect to the x-axis?
16. _____	Use $g(x) = 2x^2 - 5x$ for Questions 16 – 18. Find: 16. $g(-2)$ 17. $g(a + 2)$
17. _____	18. $\dfrac{g(a + h) - g(a)}{h}$
	19. Graph: $f(x) = 3x^2 - x - 2$.
18. _____	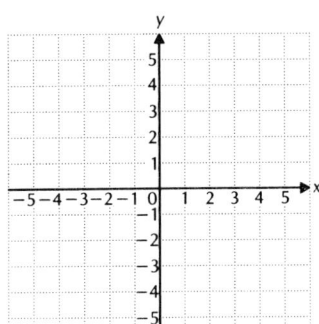
19. See graph.	
20. _____	
21. _____	Solve.
	20. $|8 - x| > 4$ 21. $x^2 + 4x - 21 > 0$
22. _____	22. $\dfrac{x + 1}{x + 2} < 2$
23. _____	23. Use synthetic division to find $P(-2)$: $$7x^4 - 3x^2 + x - 5.$$
24. _____	24. Factor the polynomial $P(x)$. Then solve the equation for $P(x) = 0$. $P(x) = x^3 + x^2 - 7x - 3$.

25. Which of the following have inverses that are functions?

a)

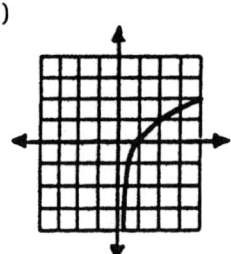

b)

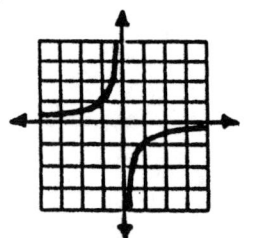

c)

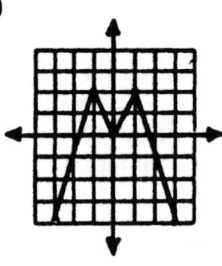

26. Simplify: $8^{\log_8 3x}$.

Solve.

27. $7^{4x-3} - 17 = 32$

28. $\log_2 (x + 1) + \log_2 (x - 1) = 3$

29. How old is an animal bone that has lost 35% of its carbon-14?

30. Students in an English class took a final exam. They took equivalent forms of the exam in monthly intervals thereafter. The average score S(t), in percent, after t months was found to be given by
$S(t) = 80 - 12 \log (t + 1)$, $t \geq 0$.

    a) What was the average score when they initially took the test, t = 0?

    b) What was the average score after 6 months?

    c) After what time was the average score 60?

ANSWERS

25. _____

26. _____

27. _____

28. _____

29. _____

30. a) _____

b) _____

c) _____

ANSWERS	
31. _____	31. A collection of 30 coins consists of nickels and quarters. The total value is $3.70. How many nickels are there?
32. _____	32. Solve using matrices. If there is more than one solution, list three of them. $$x + y + 2z = 1,$$ $$2x + y + z = 2,$$ $$3x + 2y + 3z = 3$$
33. _____	33. Evaluate: $\begin{vmatrix} \frac{5}{8} & -3 \\ 2 & 10 \end{vmatrix}$.
34. _____	34. Write a matrix equation equivalent to this system of equations and use the inverse of the coefficient matrix to solve the system. Show all your work. $$5x - y = 11,$$ $$x + 3y = -1$$
35. _____	35. Maximize and minimize T = 5x - 3y subject to: $$x + y \le 4,$$ $$-3 \le x \le 0,$$ $$y \ge 0.$$
36. _____	36. Given $C = \begin{bmatrix} 2 & -1 \\ 4 & 3 \end{bmatrix}$ and $D = \begin{bmatrix} 1 & -2 \\ -1 & 3 \end{bmatrix}$, find C + 2D.
37. _____	37. Find A^{-1}, if it exists, for $A = \begin{bmatrix} -2 & 4 \\ -1 & 3 \end{bmatrix}$.

NAME _____

	ANSWERS

38. Graph: $3x^2 + 5xy - 2y^2 = 0$.

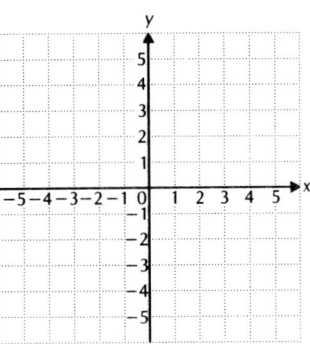

38. __See graph.__

39. _____

39. Find the vertex, focus, and directrix of the parabola $y^2 - 10y - 10x + 20 = 0$.

40. _____

40. Solve: $x^2 + 3y^2 = 21$,
    $xy = -6$.

41. _____

41. Two squares are such that the sum of their areas is 106 cm^2 and the difference of their areas is 56 cm^2. Find the length of a side of each square.

42. _____

42. Classify $\dfrac{(x - 1)^2}{49} + \dfrac{(y - 1)^2}{16} = 1$ as a circle, an ellipse, a parabola, or a hyperbola.

43. _____

43. The third term of an arithmetic sequence is 15, and the ninth term is 45. Find the eighteenth term.

ANSWERS	
	44. Insert three arithmetic means between 5 and 14.
44. _____	
	45. Find the sum of the infinite geometric sequence -10, 5, $-\dfrac{5}{2}$, $\dfrac{5}{4}$, ...
45. _____	
46. _____	
	46. Find fractional notation for $8.15\overline{15}$.
47. _____	
	47. Of the first 8 questions on a test, a student must answer 5. Of the next 6 questions, the student must answer 3. In how many ways can this be done?
48. _____	
	48. From a group of 10 men and 6 women, a committee of 5 is chosen. What is the probability that 2 men and 3 women are chosen?

Convert to scientific notation.

1. 0.0000034   2. 894.17

Simplify.

3. $(3x^2y^{-5})(-4xy^3)$ 4. $\sqrt[3]{-64}$

5. $\left(\sqrt{3} - \sqrt{2}\right)\left(\sqrt{3} + \sqrt{2}\right)$ 6. $(5a^3 - 2b^2)^2$

7. Change 80 kg/m to g/cm.

Solve.

8. $(2x - 3)(x - 1)(x + 5) = 0$ 9. $15 - 3y < 21$

10. $2(x - 5) = 7 - (x + 6)$ 11. $\dfrac{7}{x + 3} = \dfrac{5}{x + 1}$

12. Solve $\dfrac{P_1 V_1}{T_1} = \dfrac{P_2 V_2}{T_2}$ for V_2.

13. Graph:

$$f(x) = \begin{cases} x - 1, & \text{for } x \le -3 \\ x^2, & \text{for } -3 < x \le 3 \\ x - 5, & \text{for } x > 3 \end{cases}$$

ANSWERS

1. _____

2. _____

3. _____

4. _____

5. _____

6. _____

7. _____

8. _____

9. _____

10. _____

11. _____

12. _____

13. See graph.

NAME _____

ANSWERS	

14. Given $R(x) = x^2 + 15x - 340$ and $C(x) = 1.5x^2 + 3x + 20$, find $P(x)$.

14. _____

15. Find the slope and the y-intercept of the line $15 - 2x = 4y$.

15. _____

16. Find an equation of the line containing the given point and perpendicular to the given line.

$(-1,3); \; y - 2 = 4x$

16. _____

17. Find the center and the radius of the circle $x^2 + y^2 + 6x - 10y + 18 = 0$.

17. _____

18. Which of the following functions graphed below are even?

a) 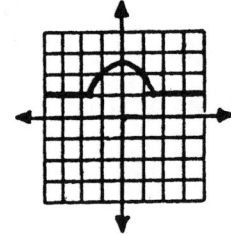　　b) 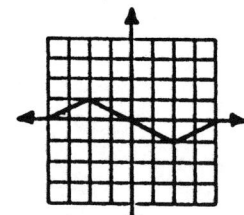　　c)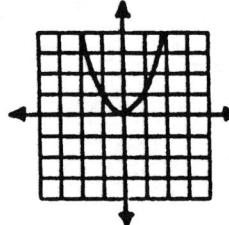

18. _____

For the functions in Questions 19 and 20:

a) Use completing the square to put each equation into the form $f(x) = a(x - h)^2 + k$;

b) find the vertex; and

c) determine whether there is a maximum or minimum function value and find that value.

19. _____

19. $f(x) = -x^2 + 5x - 6$　　　20. $f(x) = 2x^2 - 4x - 3$

20. _____

21. Find $\{2,3,5,9,12,15\} \cap \{2,4,8,12,16\}$.

21. _____

	ANSWERS
22. Use synthetic division to find the quotient and the remainder. Show all your work. $$(2x^4 - 6x^3 - x + 7) \div (x - 1).$$	22. _____
23. Find a polynomial of degree 3 having $2i$, $-2i$, and 3 as roots.	23. _____
24. What does Descartes' rule of signs tell you about the number of positive real roots and the number of negative real roots of $x^7 - x^4 - x^3 + x - 1$?	24. _____
25. Find the inverse of the relation H given by $H = \{(2,-3), (4,-2), (1,5), (-2,-7), (2.1,-1.5)\}$.	25. _____
26. Find $h(h^{-1}(-5))$: $h(x) = \dfrac{2x^3 - 5}{7}$.	26. _____
27. Express in terms of logarithms of a, b, and c: $$\log \frac{a^2 b}{c^5}.$$	27. _____
	28. _____
28. What is the loudness, in decibels, of a sound whose intensity is $8,000 \cdot I_0$?	
	29. _____
Given that $\log_{10} 2 = 0.301$, $\log_{10} 3 = 0.477$, and $\log_{10} 10 = 1$, find each of the following.	
	30. _____
29. $\log_{10} 5$ 30. $\log_{10} 9$	

ANSWERS

31. _____

31. In triangle ABC, the measure of angle B is twice the measure of angle A. The measure of angle C is $20°$ less than the measure of angle A. Find the measure of angle C.

32. _____

32. Solve using Cramer's rule: $x + 3y - z = 4,$
$x + y + 5z = -14,$
$5x - 4y + z = 2.$

33. _____

For Questions 55 - 58, let

$$A = \begin{bmatrix} 5 & -3 \\ 2 & 0 \end{bmatrix}, \quad I = \begin{bmatrix} 1 & 0 \\ 0 & 1 \end{bmatrix}, \quad \begin{matrix} F = \begin{bmatrix} 3 & 2 & -4 \end{bmatrix}, \\ 0 = \begin{bmatrix} 0 & 0 \\ 0 & 0 \end{bmatrix}, \end{matrix} \quad G = \begin{bmatrix} 1 \\ 5 \\ 0 \end{bmatrix}$$

34. _____

33. AI   34. A + 0

35. _____

35. -F·   36. FG

36. _____

37. A small bakery bakes and sells cakes and pies. To stay in business, it must sell at least 8 cakes per day, but cannot bake more than 20. It must also sell at least 16 pies, but cannot bake more than 30. It cannot bake more than 40 cakes and pies altogether. The profit on a cake is $3.50 and on a pie is $1.50. How many pies and cakes should be sold in order to maximize profit?

37. _____

	ANSWERS
38. Find the center, vertices, and foci of the ellipse $4x^2 + 9y^2 - 16x + 54y + 61 = 0$.	38. _____ _____ _____
	39. _____
39. Find an equation of the parabola with directrix $x = -4$ and focus $(4,0)$.	
	40. _____
40. The area of a rectangle is 96 ft^2, and the length of a diagonal is $4\sqrt{13}$ ft. Find the dimensions.	
	41. _____
41. Solve: $3x^2 - y^2 = 11$, $x^2 + 2y^2 = 6$.	
	42. _____
42. Classify $x^2 + 2x - 4y + 13 = 0$ as a circle, an ellipse, a parabola, or a hyperbola.	
	43. _____
43. Find the 31st term of the arithmetic sequence $-1, 6, 13, 20, \ldots$.	

ANSWERS	

44. _____

44. Evaluate the sum $\sum\limits_{k=0}^{3} (1 + 2^k)$.

45. _____

45. Find the 6th term of the geometric sequence
6, 24, 96, ...

46. _____

46. Alex was offered a job paying \$1 the first day, \$2 the
second day, and so on, doubling the salary every day
thereafter. If he takes the job, how much would he earn
in 10 days?

47. _____

47. How many distinguishable ways can the letters of the
word BEEKEEPER be arranged?

48. _____

48. Expand: $\left(x - \sqrt{3}\right)^4$.

NAME _____

CLASS _____ SCORE _____ GRADE _____

1. Write an expression containing a single radical.

$$\sqrt[3]{p} \cdot \sqrt[4]{q^3} \cdot \sqrt[5]{r^2}$$

2. Rationalize the denominator: $\dfrac{2\sqrt{x} + \sqrt{y}}{\sqrt{x} - 3\sqrt{y}}$.

3. Simplify: $\dfrac{x + \dfrac{27}{x^2}}{1 + \dfrac{3}{x}}$.

4. Write a quadratic equation whose solutions are 4 and $-\dfrac{3}{5}$.

5. Solve: $\sqrt{x + 9} = 6 - \sqrt{x - 3}$.

6. Find an equation of variation where y varies jointly as x and the square of z and inversely as w, and y = 90 when x = 18, z = 4, and w = 12.

7. If $g(x) = \dfrac{1}{2}x^2 - x + 4$, find g(−3).

8. Determine whether $y = x^3 - 3$ is symmetric with respect to the x-axis.

ANSWERS

1. _____

2. _____

3. _____

4. _____

5. _____

6. _____

7. _____

8. _____

NAME _____

9. _____

9. Find the domain: $f(x) = \dfrac{\sqrt{x - 2}}{x(x^2 - 9)}$.

10. a) _____

   b) _____

   c) _____

10. For $f(x) = 2x^2 - 6x + 1$,

   a) find an equivalent equation of the type
      $f(x) = a(x - h)^2 + k$;

   b) find the vertex; and

   c) determine whether there is a maximum or minimum,
      function value, and find that value.

11. _____

11. Solve: $|4x - 15| > 9$.

12. _____

12. Solve: $\dfrac{x - 1}{x - 2} < 2$.

13. a) _____

   b) _____

13. Given that $\log_a 2 = 0.301$, $\log_a 6 = 0.778$, and
    $\log_a 7 = 0.845$, find the following.

   a) $\log_a \sqrt{14}$ b) $\log_a 9$

14. _____

14. Find log 0.0005431.

15. _____

15. Solve: $\log_3 (\log_4 x) = 0$.

	ANSWERS

16. Solve using Cramer's Rule: $x + y - z = 6,$
$2x - y + z = 0,$
$x + 2y + z = 1.$

16. _____

17. Maximize $T = 10x + 25y$ subject to: $5x + 4y \leq 20,$
$x + y \leq 6,$
$x \geq 0,$
$y \geq 1.$

17. _____

18. Find numbers a, b, and c such that the function
$f(x) = ax^2 + bx + c$ fits the data points $(0,-4),$
$(3,-4),$ and $(2,-2).$

18. _____

19. Evaluate: $\begin{vmatrix} 2 & 0 & 3 \\ -1 & 4 & 2 \\ 3 & 1 & 5 \end{vmatrix}.$

19. _____

20. Let $A = \begin{bmatrix} 2 & 4 & 0 \\ -1 & 0 & 1 \\ 0 & -2 & 1 \end{bmatrix}$ Find A^{-1}, if it exists.

20. _____

21. Let $C = \begin{bmatrix} 1 & -2 \\ 2 & 4 \end{bmatrix}$ and $D = \begin{bmatrix} 0 & 4 \\ 3 & -1 \end{bmatrix}.$ Find $C(2D - C).$

21. _____

22. Solve: $x^2 + y^2 = 1,$
$x^2 + 25y^2 = 25.$

22. _____

NAME _____

ANSWERS	
23. _____	23. Find an equation of the circle having a diameter with endpoints $(-3,4)$ and $(3,-4)$.
24. _____	24. Find the vertices of the hyperbola: $16y^2 - 25x^2 = 400$.
25. _____	25. What does Descartes' rule of signs tell you about the number of negative real roots of $x^7 - 5x^3 - 2x + 8$?
26. _____	26. A polynomial of degree 7 with rational coefficients has 6, $-2i$, $\sqrt{5}$, and $3 - 2i$ as roots. Find the other roots.
27. _____	27. Find the rational roots of $x^4 - 5x^3 + 3x^2 + 7x - 2 = 0$.
28. _____	28. Find the sum of the first 20 terms of the arithmetic sequence 0.02, 0.05, 0.08, 0.11, ...
29. _____	29. Find the 9th term of the geometric sequence $$\frac{1}{10}, \ -\frac{1}{5}, \ \frac{2}{5}, \ -\frac{4}{5}, \ . \ . \ . \ .$$

	ANSWERS

30. Use mathematical induction. Prove for every natural number n.

$$1 + 6 + 11 + \ldots + (5n - 4) = \frac{n(5n - 3)}{2}$$

30. __See work.__

31. How many 4-digit numbers can be named using the digits 3, 4, 5, 6, 7, 8, and 9 if the digits are not repeated and the numbers must be less than 4000?

31. _____

32. Expand: $\left(2x - \sqrt{3y}\right)^4$.

33. From a group of 10 senators and 12 representatives, a committee of 8 is chosen. What is the probability that 3 senators and 5 representatives are chosen?

32. _____

Graph.

34. $f(x) = \begin{cases} x + 2, & \text{for } x < -3 \\ -x, & \text{for } -3 \le x \le 0 \\ \sqrt{x} & \text{for } x > 1 \end{cases}$

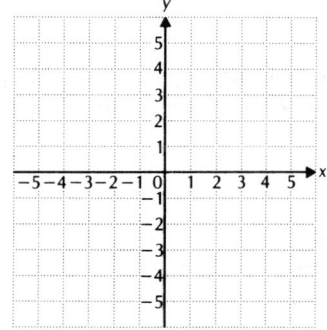

33. _____

35. $f(x) = \dfrac{x^2 - x - 1}{x - 2}$.

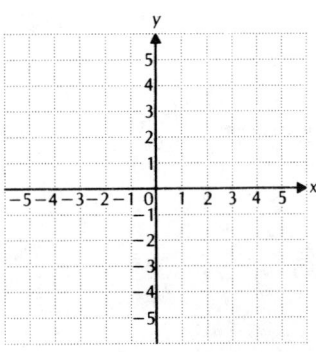

34. __See graph.__

35. __See graph.__

ANSWERS
36.   See graph.
37. _____
38. _____
39. _____
40. _____
41. _____

36. Graph: $x^2 + 8y = 0$.

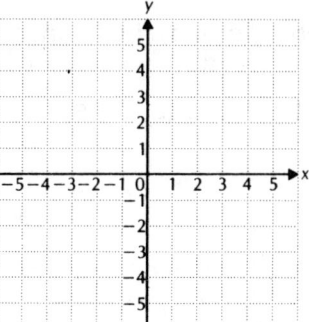

37. How many years will it take an investment of $8000 to double if interest is compounded annually at 6.5%?

38. The fourth term of an arithmetic sequence is 14, and the seventh term is 26. Find the twentieth term.

39. A rectangle has a diagonal of 26 ft and a perimeter of 68 ft. Find the dimensions.

40. Evaluate the sum $\sum_{k=1}^{5} 2(4 - 2^k)$.

41. Decompose into partial fractions: $\dfrac{7x + 5}{3x^2 + 4x + 1}$.

NAME _____

CLASS _____ SCORE _____ GRADE _____

Write the letter of your response on the answer blank.

ANSWERS

1. Subtract and simplify: $\dfrac{3x}{x^2 + 15x + 56} - \dfrac{7}{x^2 + 13x + 42}$.

   Find the numerator.

   a) $3x - 7$ b) $3x^2 + 11x - 56$

   c) $3x + 7$ d) $x - 7$

1. _____

2. Convert to scientific notation: 0.0149.

   a) 1.49×10^{-2} b) 0.149×10^{-1}

   c) 1.49×10^2 d) 149×10^3

2. _____

3. Rationalize the numerator: $\dfrac{1 - \sqrt{y}}{3 - \sqrt{y}}$.

   Find the denominator of the answer.

   a) $3 + 2\sqrt{y} - y$ b) $3 - y$

   c) $9 - y$ d) $9 - y^2$

3. _____

4. Find the center and radius of the circle:
   $x^2 + y^2 - 4x + 20y + 23 = 0$.

   a) C: $(-2,10)$, R: 3   b) C: $(2,-10)$, R: 9

   c) C: $(2,10)$, R: 9   d) C: $(2,-10)$, R: 3

4. _____

5. Solve: $4x^2 - 11x + 3 = 0$. Simplify the answer.

   a) $-\dfrac{1}{4}, 3$ b) $\dfrac{-11 \pm \sqrt{133}}{4}$

   c) $\dfrac{-11 \pm \sqrt{133}}{8}$ d) $\dfrac{11 \pm \sqrt{73}}{8}$

5. _____

6. Solve for t_2: $S = \dfrac{A}{r(t_1 - t_2)}$.

   a) $Srt_1 - A$ b) $\dfrac{Srt_1 - A}{Sr}$

   c) $t_1 - A$ d) $\dfrac{A - Srt_1}{Sr}$

6. _____

NAME _____

7. _____

7. Insert two arithmetic means between 6 and 10.

   a) $7\frac{1}{2}$, 9   b) 8, 9   c) $7\frac{1}{3}$, $8\frac{2}{3}$ d) $7\frac{2}{5}$, $9\frac{1}{5}$

8. Find $f \circ g(x)$: $f(x) = 2x^2 - 3x$, $g(x) = x + 2$.

   a) $2x^2 + 5x + 8$ b) $2x^2 - 3x + 2$

8. _____

   c) $2x^2 + 5x + 2$ d) $(2x^2 - 3x)(x + 2)$

9. Determine whether $y = 3x^3 - 2x$ is even, odd, or neither even nor odd.

   a) Even   b) Odd   c) Neither

9. _____

10. Find the domain: $g(x) = \dfrac{\sqrt{x + 1}}{x^2 - 2x}$.

    a) $\{x \mid x \neq 0, 2\}$ b) $\{x \mid x \geq -1\}$

    c) $\{x \mid x \neq 2\}$ d) $\{x \mid x \geq -1, x \neq 0, 2\}$

10. _____

11. Find an equation of the line containing the given point and parallel to the given line: $\left(\frac{1}{2}, 4\right)$; $y - 4x = 5$.

    a) $y = 4x + 2$ b) $y = 4x - 5$

11. _____

    c) $y = 4x - 4$ d) $y = \frac{x}{4} - 2$

12. Solve: $x^2 + 6x - 16 > 0$.

    a) $\{x \mid x < -8\}$ b) $\{x \mid x \neq -8, 2\}$

12. _____

    c) $\{x \mid x < -8 \text{ or } x > 2\}$ d) $\{x \mid x < -2 \text{ or } x > 8\}$

---

13. Determine whether the second coordinate of the vertex of the function $f(x) = 4x^2 - 20x + 13$ is a maximum or a minimum, and find the maximum or minimum.

    a) -12 is a minimum   b) -7 is a maximum

    c) -7 is a minimum   d) 33 is a maximum

14. Write an equivalent expression containing a single logarithm and simplify: $\log_b 3x + \log_b 10 - 2 \log_b x$.

    a) $\log_b \dfrac{30}{x}$ b) $\log_b \dfrac{5x}{10 + x^2}$

    c) 1   d) $\dfrac{\log_b 5x}{\log_b (10 + x^2)}$

15. If $f(x) = \sqrt{x - 4}$, find a formula for $f^{-1}(x)$.

    a) $f^{-1}(x) = x + 4$ b) $f^{-1}(x) = \sqrt{y - 4}$

    c) $f^{-1}(x) = x^2 - 4$ d) $f^{-1}(x) = x^2 + 4$

16. Solve: $\log_3 (x + 2) - \log_3 x = 2$.

    a) 1   b) $\dfrac{2}{7}$ ·c) $\dfrac{1}{4}$ d) 2

17. Find fractional notation for $3.1515\overline{15}$.

    a) $\dfrac{63}{20}$ b) $\dfrac{31}{11}$ c) $\dfrac{104}{33}$ d) $\dfrac{34}{11}$

ANSWERS
13. _____
14. _____
15. _____
16. _____
17. _____

ANSWERS	

18. _____

18. Solve using matrices: $x - y + z = 6$,
$2x + y + z = 3$,
$3x + 2y - z = -4$.

Find the sum of the x, y, and z values.

a) 2   b) 0   c) -2   d) -4

19. _____

19. Classify this system as consistent or inconsistent, dependent or independent.

$$5x - 6y = 4,$$
$$-10x + 12y = -6$$

a) Consistent, dependent   b) Consistent, independent

c) Inconsistent, dependent   d) Inconsistent, independent

20. _____

20. The sum of a certain number and a second number is 40. The second number minus the first number is 50. Find the smaller number.

a) 25   b) -5   c) 60   d) 45

21. _____

21. Let $A = \begin{bmatrix} 4 & -2 \\ 2 & 1 \end{bmatrix}$.   Find A^{-1} if it exists.

a) $\begin{bmatrix} 1 & 2 \\ -2 & 4 \end{bmatrix}$ b) $\begin{bmatrix} 4 & 2 \\ -2 & 1 \end{bmatrix}$ c) $\begin{bmatrix} \frac{1}{8} & \frac{1}{4} \\ -\frac{1}{4} & \frac{1}{2} \end{bmatrix}$ d) Does not exist

22. _____

22. Evaluate: $\begin{vmatrix} 3 & 0 & 4 \\ 1 & 1 & -1 \\ 2 & 0 & 2 \end{vmatrix}$.

a) -2   b) 6   c) 21   d) 14

	ANSWERS

23. Let $A = \begin{bmatrix} 1 & -2 & 3 \end{bmatrix}$ and $B = \begin{bmatrix} 3 & 0 & 2 \\ 2 & 4 & -1 \\ 0 & 2 & 1 \end{bmatrix}$.   Find AB.

23. _____

   a) $\begin{bmatrix} -1 & -2 & 7 \end{bmatrix}$ b) $\begin{bmatrix} 9 & -9 & -1 \end{bmatrix}$

   c) $\begin{bmatrix} 3 & 0 & 6 \\ 2 & -8 & -3 \\ 0 & -4 & 3 \end{bmatrix}$ d) $\begin{bmatrix} -1 \\ -2 \\ 7 \end{bmatrix}$

24. _____

24. Solve: $x^2 + y^2 = 13$,
               $3x^2 - y^2 = 3$.

   The solution set

   a) is empty.

   b) consists of one ordered pair.

   c) consists of two order pairs.

25. _____

   d) consists of four ordered pairs.

25. Find the asymptotes of the hyperbola: $x^2 - 49y^2 = 49$.

   a) $y = \frac{1}{7}x$, $y = -\frac{1}{7}x$ b) $y = 7x$, $y = -7x$

   c) $y = \frac{1}{49}x$, $y = -\frac{1}{49}x$ d) $y = 49x$, $y = -49x$

26. _____

26. A polynomial of degree 5 with rational coefficients has
   3, $1 + \sqrt{5}$, and $1 - 3i$ as roots. Find the other roots.

   a) -3, $1 - \sqrt{5}$, $1 + 3i$ b) $-\sqrt{3}$, $3i$

   c) $-1 - \sqrt{5}$, $-1 + 3i$ d) $1 - \sqrt{5}$, $1 + 3i$

27. _____

27. Given that $P(x) = 3x^5 - 4x^4 + 2$, find $P(-4)$.

   a) -4094 b) -1022 c) -4098 d) 256

28. List two of the roots of $P(x) = x^4 - x^3 - x^2 - x - 2$.

   a) -2, i b) 1, i c) $2i$, $-2i$ d) -1, i

28. _____

NAME _____

29. _____

29. The 3rd term of an arithmetic sequence is 12 and the 10th term is 61. Find the 6th term.

    a) 29   b) 7   c) 33   d) 47

30. _____

30. Find the sum of the first 6 terms of the geometric sequence $\frac{3}{4}, \frac{3}{8}, \frac{3}{16}, \ldots$

    a) $\frac{9}{32}$ b) $\frac{189}{128}$ c) $\frac{3}{256}$ d) $\frac{93}{128}$

31.   See work.

31. Use mathematical induction. Prove for every natural number n. $1 + 3 + 3^2 + \ldots + 3^{n-1} = \frac{3^n - 1}{2}$

32. _____

32. How many baseball games can be played in a 14-team league if each team plays each other team twice?

    a) 182   b) 14!   c) 1001   d) 91

33. _____

33. Find the 4th term of $(2x + y)^5$.

    a) $40x^2y^3$ b) $80x^3y^2$ c) $10xy^4$ d) $20xy^4$

34. _____

34. Suppose 3 cards are drawn without replacement from a well-shuffled deck of 52 cards. What is the probability that all three are diamonds?

    a) $\frac{3}{52}$ b) $\frac{1}{64}$ c) $\frac{1}{7225}$ d) $\frac{11}{850}$

Chapter 1, Test Form A

1. $12, 0$ 2. $\sqrt[-4]{5}, \sqrt{50}$ 3. All   4. All except $\sqrt[-4]{5}, \sqrt{50}$ 5. 12

6. $12, 0, -11$ 7. 0 8. 3.7 9. 45 10. 19 11. -3 12. $\dfrac{221}{2}$

13. 0.005401 14. $58,000$ 15. $315,000$ 16. 0.00617 17. 2.17×10^{-5}

18. 5.9321×10^2 19. 3.468×10^{-10} 20. 8.5×10^{-7} 21. $\dfrac{-10x^7}{y^8}$ 22. $\dfrac{2a^4 b^8}{3c^7}$

23. -5 24. 3 25. 4 26. $\dfrac{y - x}{(xy)(x + y)}$ 27. $16a^6 + 24a^3 b + 9b^2$

28. $7x^3 y + 4x^2 - 3xy - y^2 + 8$ 29. $8x^3 - 12x^2 + 6x - 1$ 30. $\sqrt[12]{(a + b)^5}$

31. $\dfrac{1}{\sqrt[4]{c^3}}$, or $\dfrac{\sqrt[4]{c}}{c}$ 32. $5(3x + 2)^2$ 33. $(a + 6)(a^2 - 6a + 36)$

34. $b^3(b - 2c)(b + 2c)$ 35. $2(6m^2 + 5)(m^2 - 3)$ 36. $(3 - 2a)(3 + 2a)$

$(81 + 36a^2 + 16a^4)$ 37. $(x + 2y)(5 - x)$ 38. $a^3(a^2 - 3b^2)(a^4 + 3a^2 b^2 + 9b^4)$

39. $\dfrac{x + 3}{(x + 2)^2}$ 40. $\dfrac{y^2 + 8y + 35}{(y - 3)(y + 3)(y + 7)}$ 41. $\dfrac{64 + 16\sqrt{x} + x}{64 - x}$

42. $a^{3x} + 9a^x + 27a^{-x} + 27a^{-3x}$ 43. $\dfrac{64 - x}{64 - 16\sqrt{x} + x}$ 44. $\dfrac{b^{2/3}(a - b)}{a^{1/2}}$

45. 10 46. $\dfrac{x^2(6x - 5)}{(3x - 1)^{3/2}}$ 47. $\dfrac{375}{11} \dfrac{mi}{hr}$ 48. $x^{3a} - 3x^{2a}y^b + 3x^a y^{2b} - y^{3b}$

49. $\dfrac{(3 + x)\sqrt{2 + x}}{2 + x}$

Chapter 1, Test Form B

1. $13, 0$ 2. $-\sqrt[3]{9}, \sqrt{18}$ 3. All   4. All except $-\sqrt[3]{9}, \sqrt{18}$ 5. 13

6. $13, 0, -5$ 7. $\dfrac{4}{5}$ 8. -16 9. -2.3 10. 12 11. $-\dfrac{25}{6}$ 12. $\dfrac{141}{25}$

13. $613,000$ 14. 0.04312 15. $79,200$ 16. 0.0000902 17. 3.2×10^{-6}

18. 4.37452×10^3 19. 5.022×10^{-8} 20. 4.5×10^{10} 21. $648x^7$

Chapter 1, Test Form B (continued)

22. $\dfrac{2x^3}{3yz}$ 23. -10 24. 3 25. 3 26. $\dfrac{x - y}{2x}$ 27. $25x^4 - 40x^2y^3 + 16y^6$

28. $5x^2y + 4x^3 - xy^2 - 14x - 2y - 6$ 29. $125y^3 - 300y^2 + 240y - 64$

30. $\sqrt[60]{(c + d)^{17}}$ 31. $\sqrt[5]{x^4}$ 32. $6(2x - 5)^2$ 33. $(b - 8)(b^2 + 8b + 64)$

34. $7x^3(2x^2 - z)(2x^2 + z)$ 35. $6(4a^2 - 7)(a^2 + 3)$

36. $(4 + 3a^2)(16 - 12a^2 + 9a^4)$ 37. $(2x - y)(4 - x)$

38. $a^3(a^3 - 3b^2)(a^3 + 3b^2)$ 39. $\dfrac{(x - 5)(x - 7)}{2x(x - 1)}$ 40. $\dfrac{3x + 5y}{3x - 5y}$

41. $\dfrac{16 - 8\sqrt{x} + x}{16 - x}$ 42. $27a^{3x} - 108a^{2x}b^{-x} + 144a^xb^{-2x} - 64b^{-3x}$

43. $\dfrac{16 - x}{16 + 8\sqrt{x} + x}$ 44. $\dfrac{a^{3/4}(a + b)}{b^{2/3}}$ 45. 9404.8 ft   46. $\dfrac{x^2 - 4x - 3}{(x - 3)^{3/2}}$

47. $\dfrac{27}{25}\dfrac{\text{ton}}{\text{yd}^3}$ 48. $8a^3 + 12a^2b + 36a^2c + 6ab^2 + 36abc + 54ac^2 + b^3 +$

$9b^2c + 27bc^2 + 27c^3$ 49. $\dfrac{a^3 - 3a^2b + 3ab^2}{b}$

Chapter 1, Test Form C

1. $415, 0$ 2. $-\sqrt[3]{4}, \sqrt{13}, -\sqrt{39}$ 3. All   4. All except $-\sqrt[3]{4}, \sqrt{13}, -\sqrt{39}$

5. 415 6. $415, -19, 0$ 7. -59.2 8. 5 9. 22 10. -26.39 11. $\dfrac{1}{40}$

12. $\dfrac{23}{16}$ 13. $937,000$ 14. 0.000612 15. $40,900,000$ 16. 0.0853

17. 5.37×10^{-4} 18. 8.2915×10^2 19. 5.428×10^{-7} 20. 2.5×10^{-9}

21. $625x^{12}$ 22. $\dfrac{ab^8}{4c^5}$ 23. -6 24. 4 25. $11 - 4\sqrt{7}$ 26. $\dfrac{1}{a}$

27. $6x^3 - 11x^2 + 11x - 12$ 28. $9x^4 - 3x^3 + 2x^3y - 9xy^2 + 2x - 15$

29. $64a^3 - 144a^2 + 108a - 27$ 30. $\sqrt[20]{(a - b)^{11}}$ 31. $\dfrac{1}{\sqrt[3]{a^2}}$, or $\dfrac{\sqrt[3]{a}}{a}$

$\underline{32}$. $3(5x + 1)^2$ $\underline{33}$. $(a + 9)(a^2 - 9a + 81)$ $\underline{34}$. $6x^2(y^2 - z)(y^2 + z)$

$\underline{35}$. $3(6m^2 - 1)(m^2 + 3)$ $\underline{36}$. $8(2 + 3b^2)(4 - 6b^2 + 9b^4)$ $\underline{37}$. $(x - 5y)(4 + x)$

$\underline{38}$. $a^2(a^3 + 2b^3)(a^6 - 2a^3b^3 + 4b^6)$ $\underline{39}$. $\dfrac{(x + 7)(x + 5)}{(x + 3)(x + 4)}$

$\underline{40}$. $\dfrac{3x^2 + 14x - 7}{(5x - 2)(x - 5)(x + 4)}$ $\underline{41}$. $\dfrac{25 + 10\sqrt{x} + x}{25 - x}$

$\underline{42}$. $8a^{3x} - 48a^x + 96a^{-x} - 64a^{-3x}$ $\underline{43}$. $\dfrac{25 - x}{25 - 10\sqrt{x} + x}$ $\underline{44}$. $\dfrac{a^2 - b}{a^{2/5}b^{1/2}}$

$\underline{45}$. 91.2 ft $\underline{46}$. $\dfrac{x(-2x^2 + 15)}{(x - 5)^{3/2}}$ $\underline{47}$. 240 $\dfrac{cg}{mL}$ $\underline{48}$. a^{12x}

$\underline{49}$. $(x^4 - 5)(x^4 + 5)(x^8 + 25)$

Chapter 1, Test Form D

$\underline{1}$. 0 $\underline{2}$. $\sqrt{80}$, $-\sqrt[4]{52}$ $\underline{3}$. All $\underline{4}$. All except $\sqrt{80}$, $-\sqrt[4]{52}$ $\underline{5}$. None

$\underline{6}$. -19, 0 $\underline{7}$. -14.9 $\underline{8}$. 85 $\underline{9}$. $\dfrac{23}{9}$ $\underline{10}$. -406 $\underline{11}$. 5 $\underline{12}$. $\dfrac{113}{35}$ $\underline{13}$. 0.02175

$\underline{14}$. 9100 $\underline{15}$. 5,090,000 $\underline{16}$. 0.000132 $\underline{17}$. 1.32×10^{-3} $\underline{18}$. 5.749×10^2

$\underline{19}$. 3.869×10^9 $\underline{20}$. 5.5×10^3 $\underline{21}$. $\dfrac{-12}{xy}$ $\underline{22}$. $\dfrac{c^4}{9d^6}$ $\underline{23}$. -4 $\underline{24}$. 2

$\underline{25}$. $6 + 2\sqrt{5}$ $\underline{26}$. $\dfrac{y - x}{xy(2x + y)}$ $\underline{27}$. $x^4 - a^2b^2$ $\underline{28}$. $7pq^2 + 7pq - 14q^2 - 11p +$

$19q - 9$ $\underline{29}$. $27x^3 - 54x^2 + 36x - 8$ $\underline{30}$. $\sqrt[12]{(x + y)^7}$ $\underline{31}$. $\sqrt[8]{y^3}$

$\underline{32}$. $2(3x - 4)^2$ $\underline{33}$. $(c - 10)(c^2 + 10c + 100)$ $\underline{34}$. $a^3(a - 4b)(a + 4b)$

$\underline{35}$. $2(8p^2 - 3)(p^2 + 2)$ $\underline{36}$. $(8 - 3a^3)(64 + 24a^3 + 9a^6)$ $\underline{37}$. $(x + 4y)(6 - x)$

$\underline{38}$. $b^2(2a^2 - b^2)(4a^4 + 2a^2b^2 + b^4)$ $\underline{39}$. $\dfrac{(x - 6)(x - 2)}{(x - 1)(x + 5)^2}$

$\underline{40}$. $\dfrac{y(2y + 13)}{(y - 6)(y - 1)(y + 2)}$ $\underline{41}$. $\dfrac{36 - 12\sqrt{x} + x}{36 - x}$ $\underline{42}$. $125x^{3t} + 75x^t + 15x^{-t} + x^{-3t}$

$\underline{43}$. $\dfrac{36 - x}{36 + 12\sqrt{x} + x}$ $\underline{44}$. $\dfrac{b^{5/8}(b^2 - a)}{a^{1/3}}$ $\underline{45}$. 18.4 m $\underline{46}$. $\dfrac{-4x^4 + 2x^3 + 5x^2}{(2x + 1)^{3/2}}$

Chapter 1, Test Form D (continued)

47. $400 \frac{g}{cm}$ 48. $\frac{140}{33}$ 49. $\left(y - \frac{5}{8}\right)\left(y + \frac{3}{8}\right)$

Chapter 1, Test Form E

1. $\sqrt{50}$, π, $\sqrt[3]{15}$, 17.12112111211112 . . .   2. 8   3. Associative (x)   4. 22.1

5. -11   6. 140   7. -16   8. 17   9. $-\frac{2}{9}$ 10. 0.0097   11. 5.073×10^6

12. $\frac{16x^3}{y^5}$ 13. $25a - 2$ 14. 7   15. $\frac{4s^7t^9}{5r^4}$ 16. $\frac{a^2 + b^2}{ab}$ 17. $4x^6 - 4x^3y^2 + y^4$

18. $8a^3 - 84a^2b + 294ab^2 - 343b^3$ 19. -1   20. $3x^5 - 3x^2y + 5xy - 6y^2 + 11$

21. 8.7×10^{-12} 22. $3(4x^2 + 1)(2x - 1)(2x + 1)$

23. $(y + 0.3)(y^2 - 0.3y + 0.09)$ 24. $(3x + 2)(x - 9)$

25. $(12a - 4b + 7)(6a - 2b - 3)$ 26. $\sqrt[20]{(b + 3)^7}$ 27. $\frac{1}{\sqrt[11]{z^9}}$, or $\frac{\sqrt[11]{z^2}}{z}$

28. $\frac{x^3y^4}{z^5}$ 29. $\frac{(x - 9)(x - 7)}{x(x - 2)}$ 30. $\frac{7}{a + b}$ 31. $\frac{4a - 9\sqrt{ab} + 5b}{a - b}$ 32. $\frac{3y + 5x}{x^{1/4}y^{1/3}}$

33. 4.5 m   34. $\frac{800}{3} \frac{\cent}{hr}$ 35. $y^{2b} - z^{2c}$

Chapter 1, Test Form F

1. b   2. d   3. c   4. c   5. a   6. b   7. b   8. d   9. a   10. a   11. d

12. b   13. d   14. a   15. a   16. c   17. d   18. b   19. b   20. d   21. a

22. b   23. a   24. c   25. a   26. d   27. b   28. c   29. c

Chapter 2, Test Form A

1. i   2. $-\sqrt{35}$ 3. $13 + 18i$ 4. $3 + 10i$ 5. $-\frac{7}{13} + \frac{22}{13}i$ 6. 26   7. $\frac{2}{5} + \frac{1}{5}i$

8. $x = 4$, $y = 7$ 9. $\left\{\frac{5}{3}, -7, 1\right\}$ 10. $\{\pm i, \pm 2\}$ 11. $\left\{\frac{4}{3}, 3\right\}$ 12. $\left\{\frac{1 \pm \sqrt{29}}{4}\right\}$

13. $\{2\}$ 14. $\left\{\frac{1 \pm i\sqrt{79}}{8}\right\}$ 15. $\{7\}$ 16. $\{-1\}$ 17. $\{-2, 9\}$ 18. $\{y | y > -2\}$

Chapter 2, Test Form A (continued)

19. $\{-2,2,-1\}$ 20. $\{0\}$ 21. 7 m x 21 m   22. 5 mph   23. $\{3 \pm 2\sqrt{3}\}$ 24. One real   25. $x^2 + 2\sqrt{2}x - 6 = 0$ 26. $m_1 = \dfrac{Fd^2}{km_2}$ 27. 6   28. $y = \dfrac{328xz}{w^2}$

29. 50 ft   30. No   31. $\{x \mid x \leq 5\}$ 32. $\left\{\dfrac{-3 \pm \sqrt{17}}{2}\right\}$ 33. $\left\{8 \pm 4\sqrt{3}\right\}$

Chapter 2, Test Form B

1. $-i$ 2. $-\sqrt{30}$ 3. $7 - i$ 4. $11 + 2i$ 5. $1 - 2i$ 6. 50   7. $\dfrac{5}{29} - \dfrac{2}{29}i$

8. $x = 5$, $y = -2$ 9. $\left\{9, -3, \dfrac{1}{3}\right\}$ 10. $\left\{\dfrac{1 \pm \sqrt{13}}{2}, -1, 2\right\}$ 11. $\left\{-\dfrac{5}{2}, 4\right\}$

12. $\left\{\dfrac{1 \pm \sqrt{21}}{5}\right\}$ 13. $\{1\}$ 14. $\left\{\dfrac{5 \pm i\sqrt{47}}{4}\right\}$ 15. $\{15\}$ 16. $\{2\}$ 17. $\{-5, 3\}$

18. $\{x \mid x < -2\}$ 19. $\{-3, 3, 2\}$ 20. $\{0\}$ 21. \$2203.99   22. $2\dfrac{1}{7}$ hr

23. $\left\{2 \pm \sqrt{6}\right\}$ 24. Two real   25. $x^2 - 4 = 0$ 26. $\pi = \dfrac{C}{2r}$ 27. 60 ft

28. $y = \dfrac{250xz}{w^2}$ 29. 50 cm   30. Yes   31. $\{x \mid x \geq 4\}$ 32. 56 mph   33. $\left\{-3, \dfrac{2}{k}\right\}$

Chapter 2, Test Form C

1. $-i$ 2. $\sqrt{15}$ 3. $22 - 3i$ 4. $14 + 7i$ 5. $\dfrac{21}{25} - \dfrac{22}{25}i$ 6. 10   7. $\dfrac{6}{61} + \dfrac{5}{61}i$

8. $x = 5$, $y = 2$ 9. $\left\{\dfrac{2}{5}, -4, 5\right\}$ 10. $\{16\}$ 11. $\left\{-\dfrac{2}{5}, 3\right\}$ 12. $\left\{\dfrac{2 \pm \sqrt{10}}{3}\right\}$

13. $\{10\}$ 14. $\left\{\dfrac{-1 \pm i\sqrt{31}}{4}\right\}$ 15. $\{10\}$ 16. $\{5\}$ 17. $\{-3, 6\}$ 18. $\{x \mid x > 4\}$

19. $\{-1, 1, -3\}$ 20. $\{0\}$ 21. $A = 100°$, $B = 50°$, $C = 30°$ 22. 36.5 km/h

23. $\left\{2 \pm \sqrt{6}\right\}$ 24. Two nonreal   25. $2x^2 - 5x - 12 = 0$ 26. $\ell = \dfrac{P - 2w}{2}$

27. 7 cm   28. $y = \dfrac{9}{2}xz$ 29. 3 sec   30. No   31. $\left\{x \mid x \geq \dfrac{1}{5}\right\}$ 32. $\{-2, 1\}$

33. $\left\{6 \pm 3\sqrt{3}\right\}$

Chapter 2, Test Form D

1. i 2. $\sqrt{30}$ 3. $15 + 16i$ 4. $2 - 5i$ 5. $\frac{23}{25} + \frac{14}{25}i$ 6. 17 7. $\frac{7}{58} - \frac{3}{58}i$

8. $x = 1$, $y = -1$ 9. $\left\{\frac{3}{2}, 4, 1\right\}$ 10. $\{27, -1\}$ 11. $\left\{\frac{2}{7}, -4\right\}$ 12. $\left\{\frac{5 \pm \sqrt{57}}{4}\right\}$

13. $\{2\}$ 14. $\left\{\frac{1 \pm 2i\sqrt{2}}{3}\right\}$ 15. $\{27\}$ 16. $\left\{-\frac{5}{2}, 5\right\}$ 17. $\{-6, 4\}$ 18. $\{x \mid x < 2\}$

19. $\{-4, 4, -1\}$ 20. $\{0\}$ 21. $\$1325$ 22. Bill: 3 hr; Sam: 6 hr

23. $\left\{2 \pm \sqrt{7}\right\}$ 24. Two real   25. $x^2 + 9 = 0$ 26. $V_1 = \frac{P_2 V_2 T_1}{P_1 T_2}$ 27. 15 ft,

36 ft   28. $y = \frac{225xz}{64wp}$ 29. 8.25 hr   30. Yes   31. $\left\{x \mid x \geq \frac{4}{3}\right\}$ 32. 8.26%

33. $\{86\}$

Chapter 2, Test Form E

1. h   2. p   3. a   4. e   5. i   6. n   7. r   8. w   9. z   10. t   11. s

12. o   13. c   14. q   15. v   16. $x = 7$, $y = 10$ 17. 8 hr   18. 8 km/h

19. $\$3500$ 20. $\frac{2 \pm \sqrt{3}}{2}$ 21. Two real   22. Sum: 7, Product: 3

23. $x^2 - 2x + 2 = 0$ 24. $b = \frac{a^2 + c}{2}$ 25. $c = \frac{Ab}{b - A}$ 26. $6\sqrt{2}$ 27. $y = \frac{0.2x^3}{s + w}$

28. $4\frac{2}{3}$ amps   29. $\left\{x \mid x \geq -\frac{5}{12}\right\}$ 30. No

Chapter 2, Test Form F

1. d   2. b   3. d   4. a   5. d   6. c   7. c   8. d   9. b   10. c   11. a

12. a   13. d   14. d   15. b   16. a   17. c   18. a   19. b   20. d   21. c

22. b   23. a   24. c

Chapter 3, Test Form A

1. Yes   2. {-5,1,7,8}   3. {-1,2,5,8}   4.

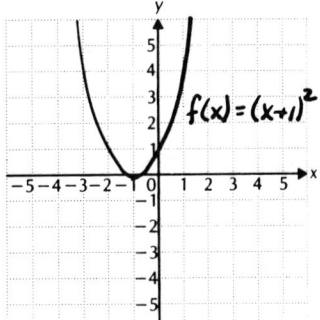

5.

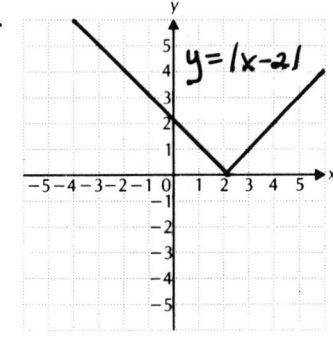

6.

7. a, d   8. c, d   9. $\{x \mid x \geq 2\}$

10. $(f + g)(x) = 2x^2 + 3x - 5$, $(f - g)(x) =$

$2x^2 - 3x + 5$, $fg(x) = 6x^3 - 10x^2$, $f/g(x) = \dfrac{2x^2}{3x - 5}$,

$f \circ g(x) = 18x^2 - 60x + 50$, $g \circ f(x) = 6x^2 - 5$

11. All reals, all reals, all reals, all reals,

all reals, all reals except $\dfrac{5}{3}$, all reals, all reals   12. b   13. 14   14. 4

15. $a^2 - 7a + 14$ 16. $2a - 3 + h$ 17. $P(x) = 1.1x^2 - 32x - 30$ 18. $m = \dfrac{7}{2}$,

y-intercept is $-\dfrac{15}{2}$ 19. $y = 2x + 9$ 20. $y = \dfrac{3}{2}x - 7$ 21. $\sqrt{20} \approx 4.472$

22. $\left(\dfrac{9}{2}, -\dfrac{1}{2}\right)$ 23. Parallel   24.

25. $y = -4x + 11$

26. $(x - 3)^2 + (y + 4)^2 = 4$

27. Center: (4,-1);
    radius: 4

28. a)

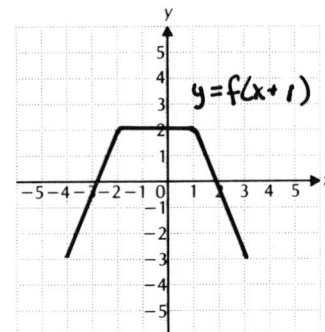

b)

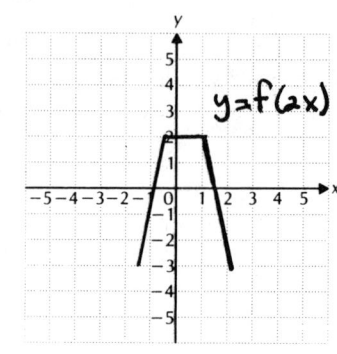

c)

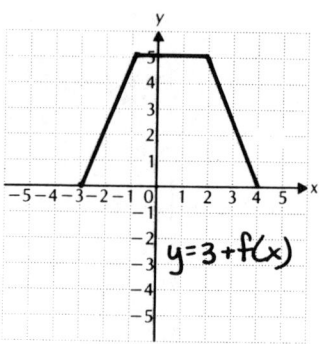

29. b   30. f   31. a, c, d, e   32. (-1,3)   33. [-7,∞)   34. b   35. a

36.

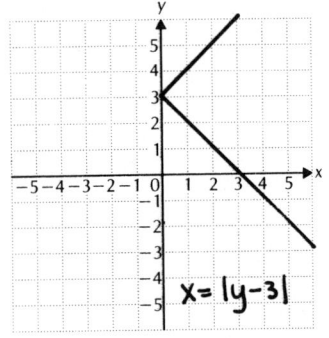

37. $A(x) = x\sqrt{144 - x^2}$ 38. (a) $f(x) = 5|x|$,

$g(x) = 3x - 2$; (b) $f(x) = 18 - 2x^2$, $g(x) = 2x - 3$;

Answers may vary.   39. $x \neq 0$, $x \neq \pm 2$

Chapter 3, Test Form B

1. No   2. {-3,1,4,5}   3. {-1,2,7}   4.

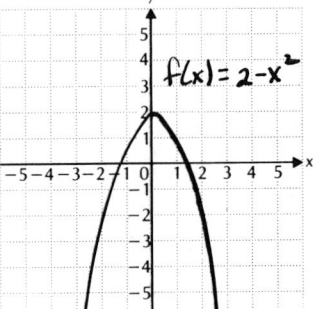

5.

6.

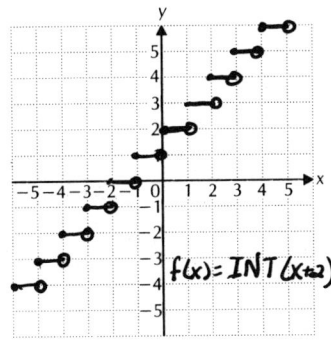

7. d   8. b, e, f   9. $\{x \mid x \neq 2\}$

10. $(f + g)(x) = x^2 + 5x + 1$, $(f - g)(x) =$

$-x^2 + 5x + 3$, $fg(x) = 5x^3 + 2x^2 - 5x - 2$,

$(f/g)(x) = \dfrac{5x + 2}{x^2 - 1}$, $f \circ g(x) = 5x^2 - 3$,

$g \circ f(x) = 25x^2 + 20x + 3$

11. All reals, all reals, all reals, all reals, all reals,

all reals except ± 1, all reals, all reals   12. a   13. -5   14. 0

15. $3a^2 + 14a + 11$ 16. $6a + 8 + 3h$ 17. $P(x) = 2.3x^2 + 37x + 290$

18. $m = -\dfrac{3}{2}$, y-intercept is 10   19. $y = -2x + 2$ 20. $y = \dfrac{1}{6}x - \dfrac{8}{3}$

21. $\sqrt{17} \approx 4.123$ 22. $\left(\dfrac{9}{2}, -3\right)$ 23. Perpendicular

24.

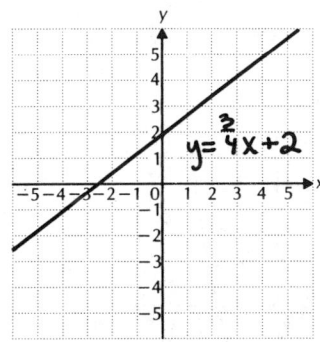

25. $y = 2x + 10$ 26. $(x + 1)^2 + (y + 3)^2 = 25$

27. Center: $(2, -5)$; radius: 5

28. a)

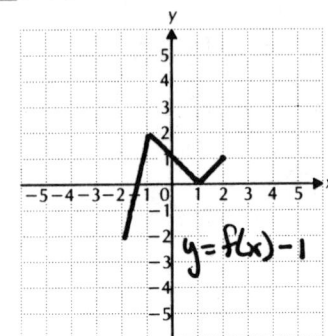

b)

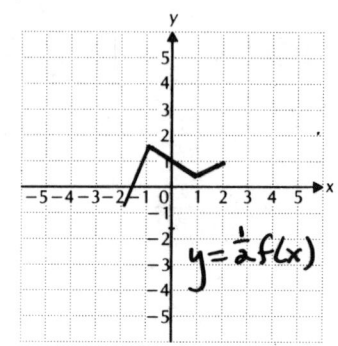

c)

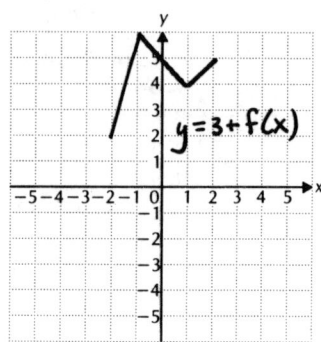

29. b, d, e   30. a, c   31. f   32. $(-\infty, -1]$ 33. $(-9, 4]$ 34. b   35. c

36.

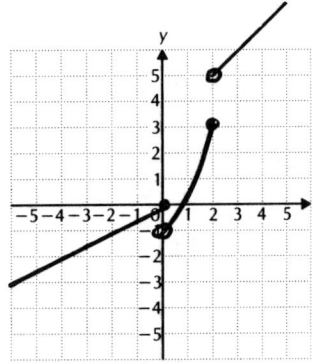

37. $A(h) = \frac{1}{2}h(2h - 4)$ 38. (a) $f(x) = \frac{1}{x^2}$,

$g(x) = x + 3$; (b) $f(x) = 5x^3 + 2$, $g(x) = x - 3$;

Answers may vary.   39. $k = -4$

Chapter 3, Test Form C

1. Yes   2. $\{-6, -1, 3, 5\}$ 3. $\{-1, 0, 3, 5\}$ 4.

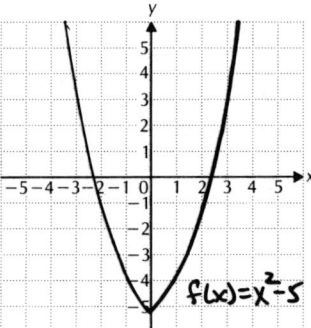

5.

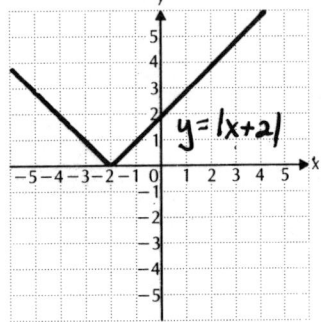

6.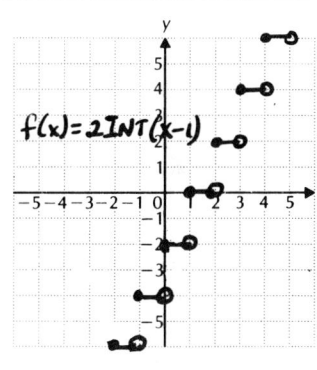

$f(x) = 2\,INT(x-1)$

7. d,f   8. f   9. $\{x \,|\, x \neq 0\}$

10. $(f + g)(x) = x^2 - 4x + 12$, $(f - g)(x) =$

$6 - 4x - x^2$, $fg(x) = -4x^3 + 9x^2 - 12x + 27$,

$(f/g)(x) = \dfrac{9 - 4x}{x^2 + 3}$, $f \circ g(x) = -4x^2 - 3$,

$g \circ f(x) = 16x^2 - 72x + 84$

11. All reals, all reals, all reals, all reals, all reals,

all reals, all reals, all reals   12. (b)   13. -7   14. -1   15. $b^2 + 9b + 13$

16. $2a + 5 + h$ 17. $P(x) = 0.5x^2 + 64x - 25$ 18. $m = \dfrac{2}{5}$, y-intercept is $-\dfrac{9}{5}$

19. $y = 2x - 7$ 20. $y = -\dfrac{7}{5}x + \dfrac{18}{5}$ 21. $\sqrt{85} \approx 9.220$ 22. $\left(\dfrac{1}{2}, -\dfrac{5}{2}\right)$

23. Neither   24.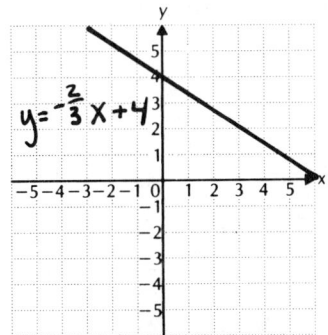

$y = -\dfrac{2}{3}X + 4$

25. $y = -2x - 5$

26. $(x + 1)^2 + (y - 5)^2 = 9$ 27. Center: (3,-2); radius: $\sqrt{5}$

<u>28</u>. a)

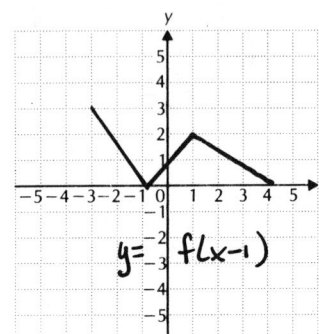

b)

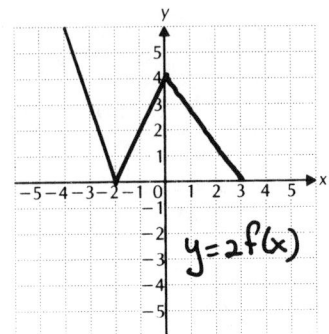

c)

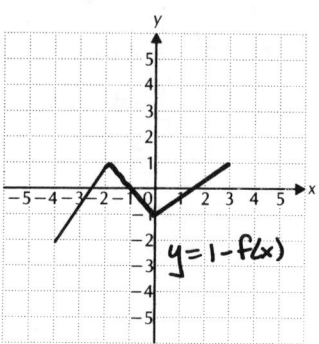

<u>29</u>. b, f   <u>30</u>. c, e   <u>31</u>. a, d   <u>32</u>. (-2,5]   <u>33</u>. (-∞,-4)   <u>34</u>. (a)   <u>35</u>. (b)

and (c)   <u>36</u>.

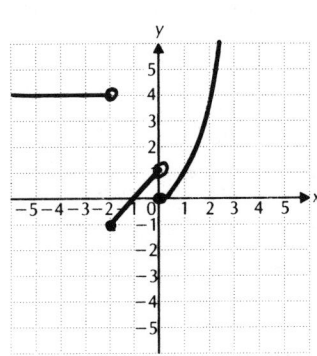

<u>37</u>. A(x) = x(10 - x)

<u>38</u>. (a) $f(x) = \sqrt[3]{x}$, g(x) = 3x - 4;

(b) $f(x) = 2x^2 + x + 7$, g(x) = x - 4;

Answers may vary.   <u>39</u>. (3,0)

Chapter 3, Test Form D

<u>1</u>. No   <u>2</u>. {-1,3,4,5}   <u>3</u>. {-1,3,5}   <u>4</u>.

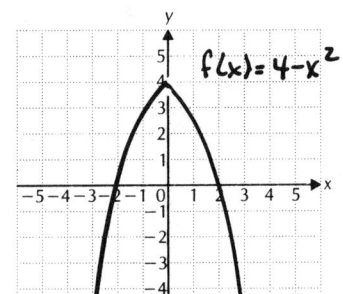

<u>5</u>.

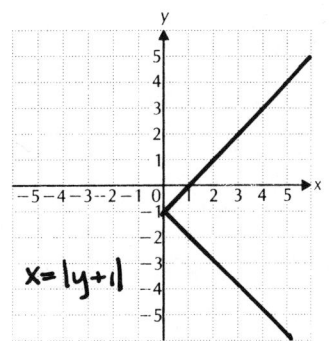

6.

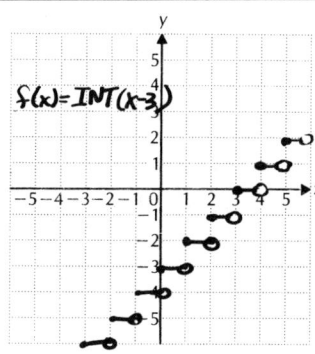

$f(x) = INT(x-3)$

7. d, f   8. none   9. $\{x \mid x \geq -\frac{3}{2}\}$

10. $(f + g)(x) = 3x^2 + 2x + 3$, $(f - g)(x) =$ $3x^2 - 2x - 5$, $fg(x) = 6x^3 + 12x^2 - 2x - 4$

$(f/g)(x) = \frac{3x^2 - 1}{2x + 4}$, $f \circ g(x) = 12x^2 + 48x + 47$,

$g \circ f(x) = 6x^2 + 2$

11. All reals, all reals, all reals, all reals, all reals, all reals except -2, all reals, all reals   12. a   13. 52   14. 0   15. $2a^2 - 9a + 7$

16. $4a - 5 + 2h$ 17. $P(x) = 1.1x^2 + 18x - 29$ 18. $m = -\frac{2}{5}$, y-intercept is 3

19. $y = -3x + 1$ 20. $y = \frac{4}{7}x + \frac{23}{7}$ 21. $\sqrt{89} \approx 9.434$ 22. $\left(4, -\frac{1}{2}\right)$

23. Perpendicular   24.

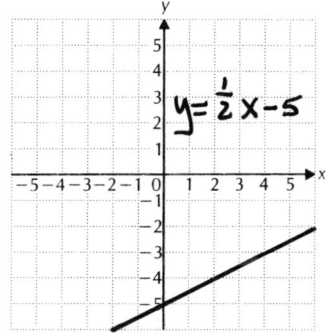

$y = \frac{1}{2}x - 5$

25. $y = \frac{1}{2}x - \frac{7}{2}$ 26. $(x - 2)^2 + (y + 6)^2 = 16$ 27. Center: (1,-6); radius: 6

28. a)

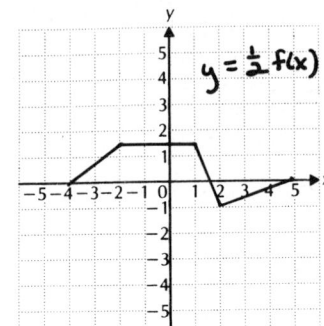

b)

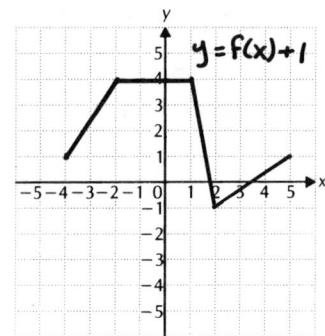

c)

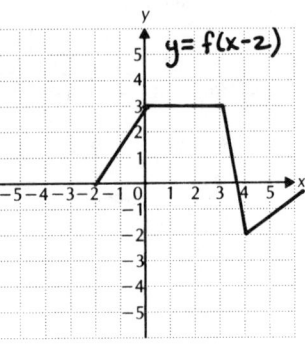

29. b, f   30. a, c   31. d, e   32. (-5,2]   33. (-∞,6]   34. c   35. None

36.

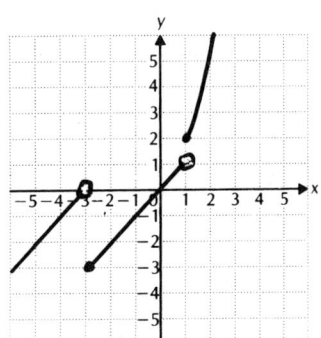

37. $A(x) = x\sqrt{400 - x^2}$

38. (a) $f(x) = \dfrac{1}{\sqrt{x}}$, $g(x) = 2x - 1$

(b) $f(x) = 6x^2 + 5$, $g(x) = x + 2$ 39. k = -13;

Answers may vary.

Chapter 3, Test Form E

1. Yes   2. {-5,-2,0,3,5}   3. {-1,2,4,5}   4.

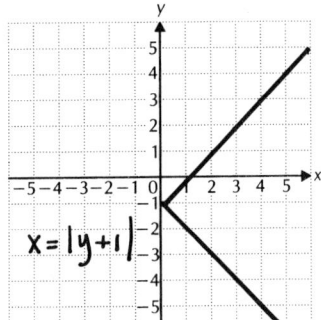

5.

$y = x^2 + 2$

6.

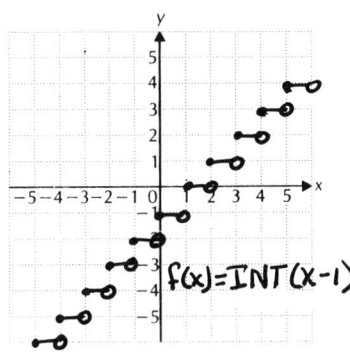

$f(x) = INT(x-1)$

7. f   8. a, d, f   9. c, f   10. $\{x | x \neq -1 \text{ or } x \neq 4\}$

11. $(f + g)(x) = 4x^2 + 2x - 4$, $(f - g)(x) = -4x^2 + 2x - 2$, $fg(x) = 8x^3 - 12x^2 - 2x + 3$,

$(f/g)(x) = \dfrac{2x - 3}{4x^2 - 1}$, $f \circ g(x) = 8x^2 - 5$,

$g \circ f(x) = 16x^2 - 48x + 35$ 12. All reals,

all reals, all reals, all reals, all reals, all reals except $-\dfrac{1}{2}$, $\dfrac{1}{2}$, all reals,

all reals   13. b, c, d   14. 49   15. 35   16. $3a^2 - 13a + 19$

17. $6a - 1 + 3h$ 18. $P(x) = -0.2x^2 + 27x + 2$

19. (a)

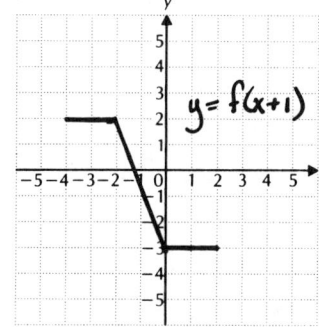

$y = f(x+1)$

(b)

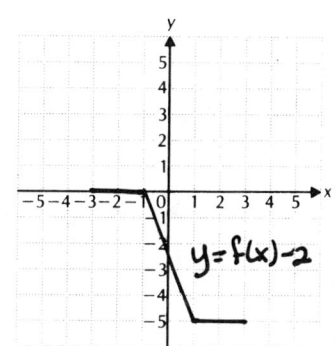

$y = f(x) - 2$

(c)

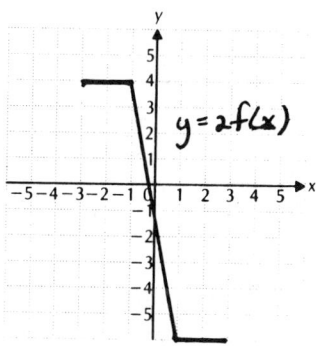

$y = 2f(x)$

20. c, e   21. a, b, f   22. d   23. $[-2, 5)$ 24. $(-3, \infty)$ 25. a

26. b   27. m = -6, y-intercept is 10   28. y = 5x - 7   29. $\sqrt{10} \approx 3.162$

Chapter 3, Test Form E (continued)

30.

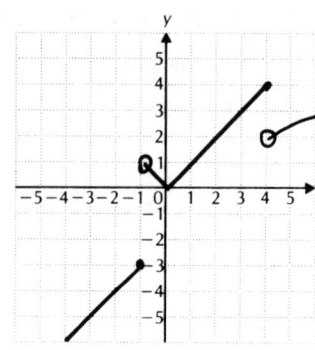

31. $A = t\sqrt{256 - t^2}$

32. $y = -\frac{2}{3}x + \frac{10}{3}$

33. $f(x) = \frac{5\sqrt{x}}{2}$, $g(x) = 3x - 8$

34. $(x + 1)^2 + (y - 5)^2 = 144$

Chapter 3, Test Form F

1. b   2. d   3. b   4. c   5. b   6. c   7. a   8. a   9. d   10. c   11. d

12. b   13. b   14. c   15. d   16. b   17. b   18. b

Chapter 4, Test Form A

1. (a) $f(x) = 2(x - 2)^2 - 7$; (b) $(2,-7)$; (c) minimum: -7

2. (a) $f(x) = -3\left(x - \frac{1}{3}\right)^2 - \frac{2}{3}$; (b) $\left(\frac{1}{3}, -\frac{2}{3}\right)$; (c) maximum: $-\frac{2}{3}$

3.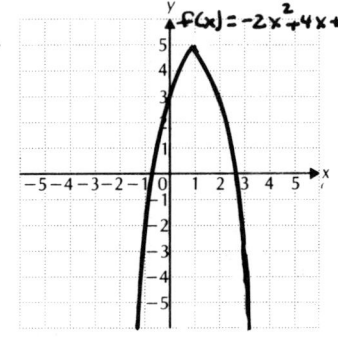

4. $\left(\frac{3 - \sqrt{41}}{8}, 0\right)$, $\left(\frac{3 + \sqrt{41}}{8}, 0\right)$ 5. $\{1,2,5,7,9,11\}$

6. ⟵├┼┼┼┼┼┼┼○┼●┼┼⟶   7. $(-8,-1]$
       0 1 2 3 4 5

8. $(-\infty,-1] \cup [6,\infty)$ 9. $(4,10)$ 10. $\{-2,\frac{10}{3}\}$

11. $(\infty,1) \cup (9,\infty)$ 12. $\left(-1,\frac{2}{3}\right]$

13. $(-\infty,-11) \cup (-4,\infty)$ 14. $-8, -8$

15.

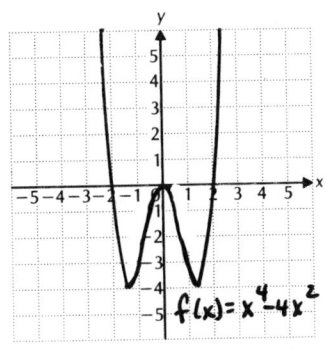

16.

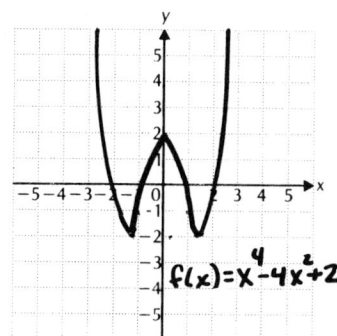

17. 499   18. Yes   19. The quotient is $4x^2 + x + 3$. The remainder is 1.

20. 100   21. $(x - 4)(x - 1)(x + 3)$; 4, 1, -3 22. $x^4 - 2x^3 - 7x^2 + 18x - 18$

23. $(x - 2)(x + 3)(x - 1)^2(x - 5)^3$ 24. $x^3 - 11x^2 + 41x - 51$

25. $\pm\left(\dfrac{1}{4}, \dfrac{1}{2}, \dfrac{3}{4}, 1, \dfrac{3}{2}, 2, 3, 6\right)$ 26. Positive: 3 or 1; Negative: 1

27. $[-1,2]$ 28. ± 2.3 29.

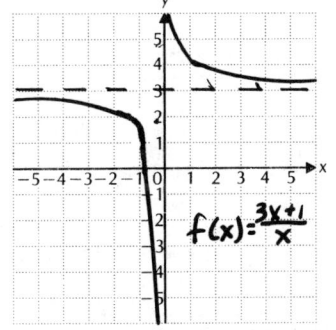

30. $k = \dfrac{5}{6}$

31. $\left(-\infty, -\dfrac{3}{2}\right]$

32. $-2 < x < 3$ or $x > 4$

Chapter 4, Test Form B

1. (a) $f(x) = -3(x - 2)^2 + 8$; (b) $(2,8)$; (c) maximum: 8

2. (a) $f(x) = 5\left(x + \dfrac{1}{5}\right)^2 + \dfrac{4}{5}$; (b) $\left(-\dfrac{1}{5}, \dfrac{4}{5}\right)$; (c) minimum: $\dfrac{4}{5}$

3.

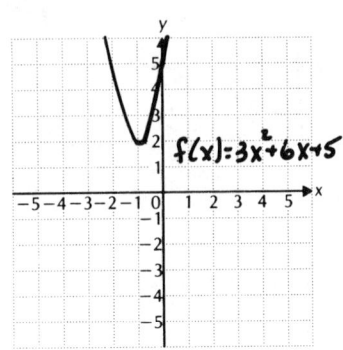

4. $\left(\dfrac{-5 - \sqrt{97}}{4}, 0\right)$, $\left(\dfrac{-5 + \sqrt{97}}{4}, 0\right)$ 5. $\{5, 15\}$

6.

7. ∅

8. $(-\infty, -1) \cup \left(\dfrac{5}{3}, \infty\right)$ 9. $[-2, 3]$ 10. $\{-4, \dfrac{22}{5}\}$

11. $(-\infty, -4) \cup (2, \infty)$ 12. $\left(-\dfrac{1}{2}, \dfrac{1}{5}\right)$ 13. $\left(-\dfrac{3}{2}, -\dfrac{2}{3}\right)$

14. $-9, 9$

15.

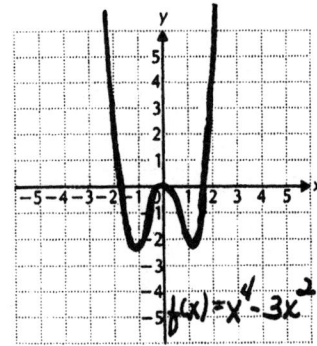

16.

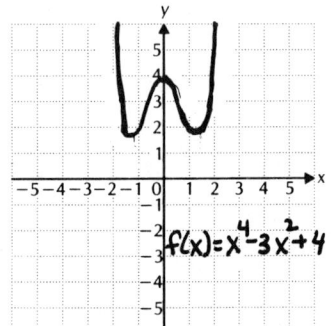

17. 4   18. Yes   19. The quotient is $3x^3 - 6x^2 + 10x - 17$.

The remainder is 33.   20. 323   21. $(x - 5)(x + 1)(x + 2)$; 5, −1, −2

22. $x^3 - 5x^2 - 3x + 15$ 23. $x(x - 5)(x - 3)^2(x + 1)^4$

24. $x^3 - 13x^2 + 49x - 49$ 25. $\pm \left(\dfrac{1}{2}, 1, 2, 4, 8\right)$ 26. Positive: 3 or 1;

Negative: 3 or 1   27. $[-2, 1]$ 28. 1.8

Chapter 4, Test Form B (continued)

<u>29</u>.

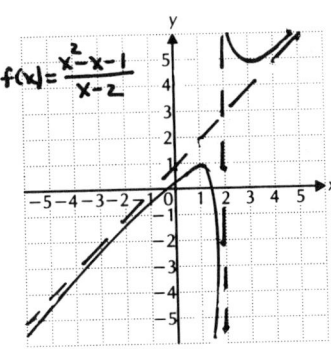

$f(x) = \dfrac{x^2 - x - 1}{x - 2}$

<u>30</u>. 36   <u>31</u>. $\left[-\sqrt{5}, \sqrt{5}\right]$ <u>32</u>. k = -13

Chapter 4, Test Form C

<u>1</u>. (a) $f(x) = 6(x + 1)^2 - 11$; (b) (-1,-11); (c) minimum: -11

<u>2</u>. (a) $f(x) = -2\left(x - \dfrac{1}{4}\right)^2 + \dfrac{25}{8}$; (b) $\left(\dfrac{1}{4}, \dfrac{25}{8}\right)$; (c) maximum: $\dfrac{25}{8}$

<u>3</u>.

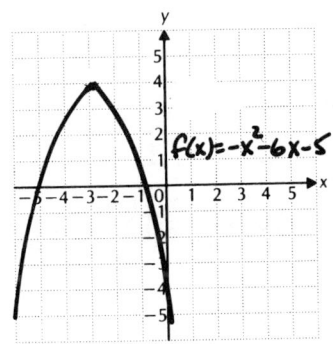

$f(x) = -x^2 - 6x - 5$

<u>4</u>. $\left(\dfrac{7 - \sqrt{61}}{6}, 0\right)$, $\left(\dfrac{7 + \sqrt{61}}{6}, 0\right)$

<u>5</u>. {1,4,7,8,9,11,16,18,20}

<u>6</u>. ⟵|┼┼┼┼●┼┼┼○┼┼⟶
          0 1 2 3 4 5

<u>7</u>. [-1,1)

<u>8</u>. $\left(-\infty, -\dfrac{4}{5}\right] \cup [2,\infty)$ <u>9</u>. (-16,-8)   <u>10</u>. $\left\{-3, \dfrac{15}{4}\right\}$

<u>11</u>. $(-\infty, -6) \cup (2,\infty)$ <u>12</u>. $\left(-\dfrac{1}{4}, \dfrac{2}{3}\right)$ <u>13</u>. (-4,-1)

<u>14</u>. Base: 14 cm; height: 14 cm

15.

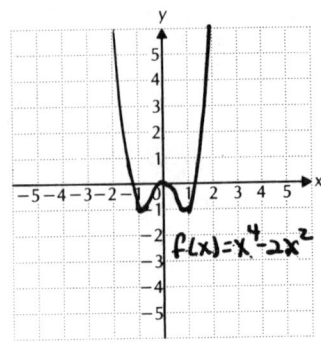

16.

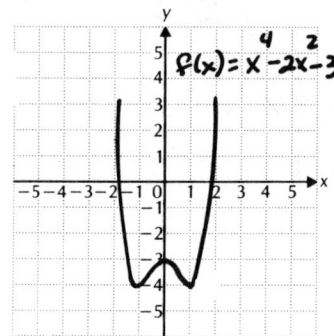

17. 57   18. No   19. The quotient is $5x^2 + 13x + 55$.   The remainder is

216.   20. 421   21. $(x - 4)(x - 2)(x + 7)$; 4, 2, -7   22. $x^4 - 7x^2 - 144$

23. $(x + 4)(x + 3)(x - 5)^2(x + 2)^3$ 24. $x^3 + 8x^2 + 46x + 68$

25. $\pm \left(\frac{1}{6}, \frac{1}{3}, \frac{1}{2}, \frac{2}{3}, 1, \frac{4}{3}, 2, 4 \right)$ 26. Positive: 2 or 0; Negative: 3 or 1

27. [-2,1]   28. -1.2, 1.4   29.

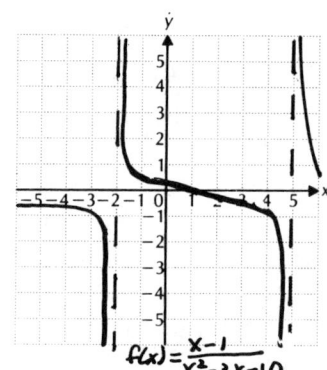

30. $\{x \mid x > -4\}$, or $(-4, \infty)$

31. $\{h \mid h > -3 + \sqrt{39} \text{ cm}\}$

32. $k = -\frac{3}{2}$

Chapter 4, Test Form D

1. (a) $f(x) = -2(x + 2)^2 + 11$; (b) (-2,11); (c) maximum: 11

2. (a) $f(x) = 4\left(x + \frac{5}{8}\right)^2 - \frac{57}{16}$; (b) $\left(-\frac{5}{8}, -\frac{57}{16}\right)$; (c) minimum: $-\frac{57}{16}$

3.

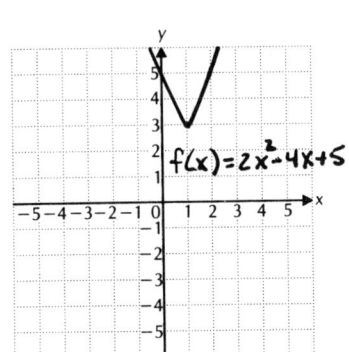

4. $\left(\dfrac{-5 - \sqrt{109}}{6}, 0\right)$, $\left(\dfrac{-5 + \sqrt{109}}{6}, 0\right)$ 5. $\{1, 7\}$

6. 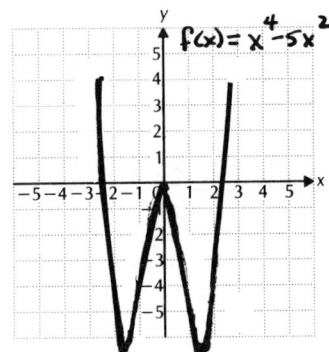   7. $(-2, 3)$

8. $(-\infty, -2) \cup (16, \infty)$ 9. $\left[-\dfrac{7}{3}, 1\right]$ 10. $\{-4, \dfrac{3}{2}\}$

11. $(-\infty, -4) \cup (6, \infty)$ 12. $\left(-\dfrac{1}{7}, 4\right]$ 13. $(-7, 8]$

14. -11, 11

15.

16.

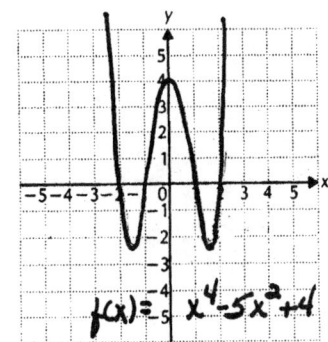

17. 225   18. No   19. The quotient is $x^3 - x^2 + 3x - 10$. The remainder is

34. 20. 2246   21. $(x - 5)(x - 3)(x + 1)$; 5, 3, -1 22. $x^3 - 11x^2 + 27x + 7$

23. $(x - 1)(x + 2)(x + 1)^2(x - 6)^3$ 24. $x^3 - 6x^2 + 4x + 16$

25. $\pm \left(\dfrac{1}{8}, \dfrac{1}{4}, \dfrac{1}{2}, 1, 2, 4\right)$ 26. Positive: 4, 2, or 0; Negative: 0   27. $[-1, 2]$

28. 0.9

29.

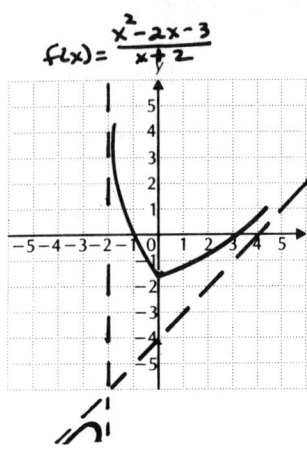

$f(x) = \dfrac{x^2 - 2x - 3}{x + 2}$

30. $\{x \mid x > 5\}$, or $(5, \infty)$ 31. (a) 10, 22;

(b) $\{x \mid 10 < x < 22\}$, or $(10, 22)$ 32. $k = -7$

Chapter 4, Test Form E

1. (a) $f(x) = 2\left(x - \dfrac{5}{2}\right)^2 - \dfrac{35}{2}$; (b) $\left(\dfrac{5}{2}, -\dfrac{35}{2}\right)$; (c) minimum: $-\dfrac{35}{2}$

2.

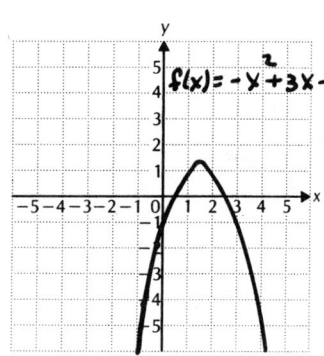

$f(x) = -x^2 + 3x - 1$

The x-intercepts are $\left(\dfrac{3 - \sqrt{5}}{2}, 0\right)$ and $\left(\dfrac{3 + \sqrt{5}}{2}, 0\right)$

3. $\ell = 27$, $w = 27$ 4. $\{b, d, f, h, j, \ell, p\}$

5. ⟵┼┼┼┼●●┼┼┼┼┼┼⟶
   -5 -4 -3 -2 -1 0 1 2 3 4 5   6. j   7. h   8. e

9. f   10. a   11. b   12. g

13.

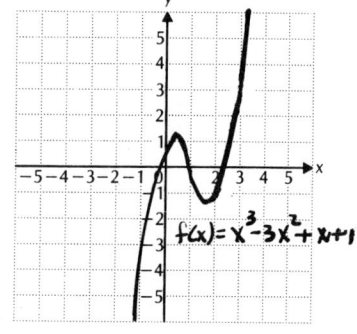

$f(x) = x^3 - 3x^2 + x + 1$

14.

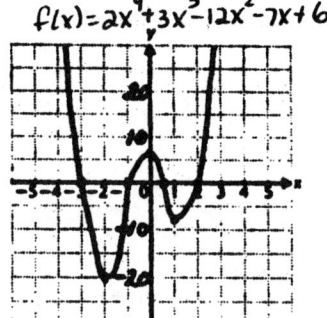

$f(x) = 2x^4 + 3x^3 - 12x^2 - 7x + 6$

Chapter 4, Test Form E (continued)

<u>15</u>. 8   <u>16</u>. The quotient is $5x^4 - 10x^3 + 18x^2 - 35x + 68$. The remainder is

-133.   <u>17</u>. 834   <u>18</u>. $(x - 5)(x - 1)(x + 1)(x + 3)(x + 5)$; 5, 1, -1, -3, -5

<u>19</u>. $x^4 - 2x^3 + 2x^2 - 8x - 8$ <u>20</u>. $x(x + 2)(x - 5)^4(x + 7)^3$ <u>21</u>. $-2 \pm 2i\sqrt{3}$

<u>22</u>. $2i$, $-\sqrt{7}$, $4 + 5i$ <u>23</u>. $\pm\left(\dfrac{1}{6}, \dfrac{1}{3}, \dfrac{1}{2}, \dfrac{2}{3}, 1, \dfrac{4}{3}, 2, 4\right)$ <u>24</u>. 3 or 1   <u>25</u>. $[-2,2]$

<u>26</u>.

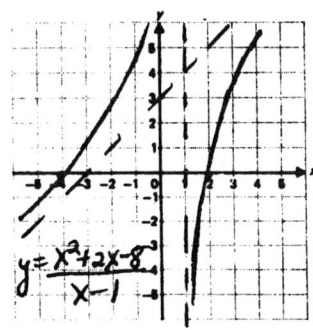

Chapter 4, Test Form F

<u>1</u>. d   <u>2</u>. b   <u>3</u>. a   <u>4</u>. a   <u>5</u>. b   <u>6</u>. a   <u>7</u>. c   <u>8</u>. d   <u>9</u>. c   <u>10</u>. b   <u>11</u>. a

<u>12</u>. d   <u>13</u>. c   <u>14</u>. b   <u>15</u>., c   <u>16</u>. a   <u>17</u>. d   <u>18</u>. b   <u>19</u>. c   <u>20</u>. a

Chapter 5, Test Form A

<u>1</u>. {(-1,2), (-3,5), (1,7), (2,-4), (4.5,3.2)}   <u>2</u>. x = 3y - 3   <u>3</u>. b

<u>4</u>. $f^{-1}(x) = \dfrac{6 - x}{2}$ <u>5</u>. 3   <u>6</u>.

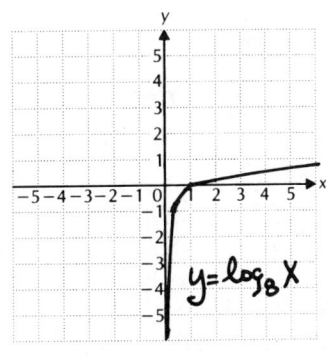

<u>7</u>.

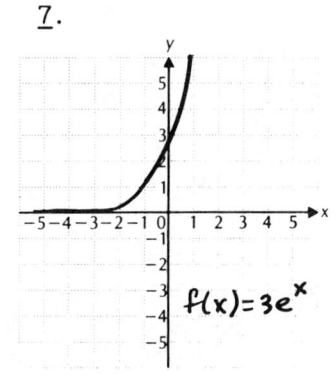

$\underline{8}$. 2.5 $\underline{9}$. $\left(\sqrt{3}\right)^6 = 27$ $\underline{10}$. $\log_x 0.2401 = 4$ $\underline{11}$. $x + 2$ $\underline{12}$. $\log_a \dfrac{\sqrt[4]{x^3 z^5}}{\sqrt{y}}$

$\underline{13}$. $2 \log a - \log b$ $\underline{14}$. 0.477 $\underline{15}$. 1.447 $\underline{16}$. 0.1945 $\underline{17}$. $-\sqrt{2}, \sqrt{2}$

$\underline{18}$. 64 $\underline{19}$. $\dfrac{1}{2}$ $\underline{20}$. 1 $\underline{21}$. 0.9163 $\underline{22}$. Not defined $\underline{23}$. -11.2506

$\underline{24}$. 0.7178 $\underline{25}$. -1.3593 $\underline{26}$. 4.6946 $\underline{27}$. 7.3 yr $\underline{28}$. 6.34 $\underline{29}$. 106,666

$\underline{30}$. (a) 2.4 ft/sec; (b) 3,801,851 $\underline{31}$. (a) $P(t) = 0.15e^{0.084t}$; (b) \$1.22;

(c) 2005; (d) about 8 years $\underline{32}$. 3.5% $\underline{33}$. 10.2 yr $\underline{34}$. 4540 yr

$\underline{35}$. 27 decibels $\underline{36}$. 4.6 $\underline{37}$. $\pi^{4.2}$ $\underline{38}$. $\dfrac{11}{3}$

Chapter 5, Test Form B

$\underline{1}$. {(-2,8), (-7,6), (3,8), (1,-5), (-1.2,5.6)} $\underline{2}$. $x = y^2 + 1$ $\underline{3}$. b

$\underline{4}$. $f^{-1}(x) = \dfrac{3 - 2x}{x}$ $\underline{5}$. -2 $\underline{6}$.

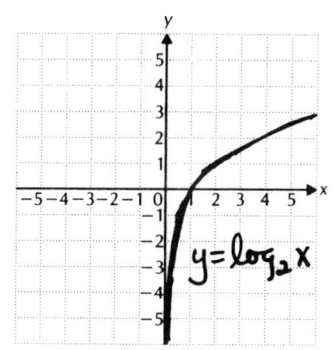

$\underline{7}$.

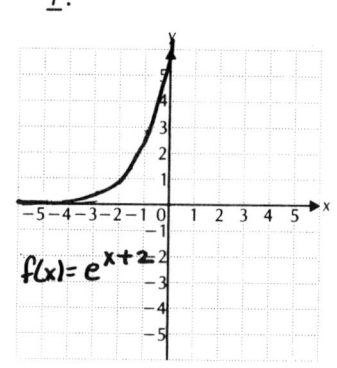

$\underline{8}$. 1.7604 $\underline{9}$. $\left(\sqrt{2}\right)^8 = 16$ $\underline{10}$. $\log_x 0.00243 = 5$ $\underline{11}$. $4x$ $\underline{12}$. $\log_a \dfrac{\sqrt{xz^4}}{y^3}$

$\underline{13}$. $2 \log x + \dfrac{1}{2} \log y$ $\underline{14}$. 0.602 $\underline{15}$. 2.301 $\underline{16}$. 0.1505 $\underline{17}$. -3, 3 $\underline{18}$. 36

$\underline{19}$. 1 $\underline{20}$. 2 $\underline{21}$. 0.1823 $\underline{22}$. 4.4048 $\underline{23}$. 1.5933 $\underline{24}$. -9.8837

$\underline{25}$. -2.4895 $\underline{26}$. Not defined $\underline{27}$. 11.9 yr $\underline{28}$. 38 decibels $\underline{29}$. 112,500

<u>30</u>. (a) 3000; (b) 3180; (c) \$100,000,000   <u>31</u>. (a) $P(t) = 4.5e^{0.032t}$;

(b) 7.5 million; (c) 1993   <u>32</u>. 2.0%   <u>33</u>. 7.4 yr   <u>34</u>. 7229 yr

<u>35</u>. 2.54   <u>36</u>. 7.9×10^{-7} <u>37</u>. (a) 125.000000; (b) 156.590645; (c) 156.842871;

(d) 156.969136   <u>38</u>. 4

Chapter 5, Test Form C

<u>1</u>. {(3,−1), (−7,4), (0,6), (5,2), (7.6,−1.3)}   <u>2</u>. $x = |y + 3|$ <u>3</u>. a, b

<u>4</u>. $f^{-1}(x) = \dfrac{x + 2}{x - 1}$ <u>5</u>. −2   <u>6</u>.   <u>7</u>.

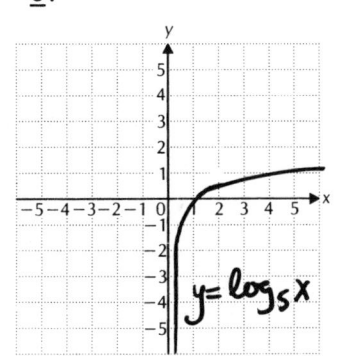

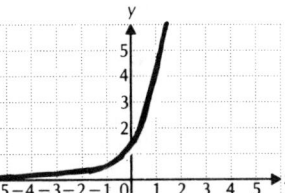

<u>8</u>. 3.3802   <u>9</u>. $\left(\sqrt{5}\right)^4 = 25$ <u>10</u>. $\log_x 0.015625 = 6$ <u>11</u>. $x - 1$ <u>12</u>. $\log_a \dfrac{x^5 \sqrt[3]{y}}{z^2}$

<u>13</u>. $\log 5 + \log c + 2 \log d$ <u>14</u>. 0.301   <u>15</u>. 1.681   <u>16</u>. 0.259   <u>17</u>. −2, 2

<u>18</u>. 125   <u>19</u>. $\dfrac{4}{5}$ <u>20</u>. 9   <u>21</u>. −0.1054   <u>22</u>. −13.0736   <u>23</u>. 5.0052   <u>24</u>. −0.8625

<u>25</u>. Not defined   <u>26</u>. 1.6332   <u>27</u>. 9.0 yr   <u>28</u>. 5.17   <u>29</u>. 285,610

<u>30</u>. (a) 1.3 yr; (b) 5.8 yr   <u>31</u>. (a) $P(t) = 0.03e^{0.038t}$; (b) 58¢;

(c) 2035 years   <u>32</u>. 3.8%   <u>33</u>. 6.9 yr   <u>34</u>. 3590 yr   <u>35</u>. 36 decibels

<u>36</u>. 1.3×10^{-5} <u>37</u>. 16   <u>38</u>. −16, 16

Chapter 5, Test Form D

1. {(3,-4), (-5,2), (4,1), (4.3,6.2), (7,0)}   2. $yx = 3$ 3. a, b

4. $f^{-1}(x) = x^2 - 3$ 5. -7   6.   7.

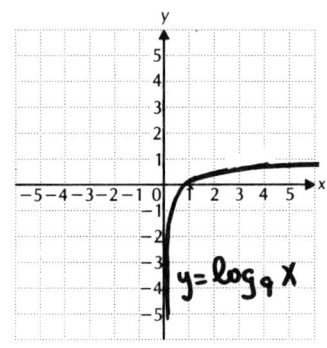

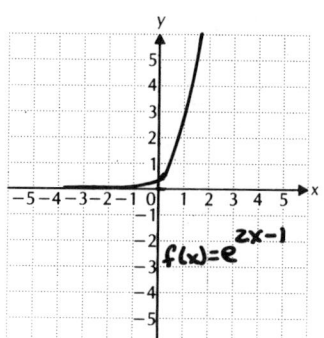

8. 4.6439   9. $\left(\sqrt{5}\right)^6 = 125$ 10. $\log_x 0.00032 = 5$ 11. 5x   12. $\log_a \dfrac{x^2 y^4}{\sqrt[3]{z^2}}$

13. $\dfrac{2}{3} \log a + \dfrac{1}{3} \log b$ 14. 0.477   15. 2   16. 0.5395   17. -1, 1   18. 8

19. 3   20. 8   21. 0.3567   22. 0.4615   23. 1.8294   24. 4.2393   25. Not

defined   26. -9.2926   27. 6.6 yr   28. 33 decibels   29. 104,509

30. (a) 75; (b) 63; (c) 60 months   31. (a) $P(t) = 100e^{0.182t}$; (b) 617;

(c) 3.8 days   32. 1.9%   33. 8.0 yr   34. 2287 yr   35. 3.91   36. 5.7

37. $\left(\sqrt{7}\right)^{-3}$, or $7^{-3/2}$ 38. $\{x \mid x > e^{4/3}\}$

Chapter 5, Test Form E

1. a, b   2. $f(x)^{-1} = \sqrt{x - 3}$ 3. $f^{-1}(x) = (4x - 20)^2$ 4. 6

5.

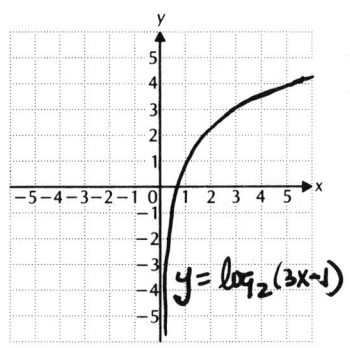

6.

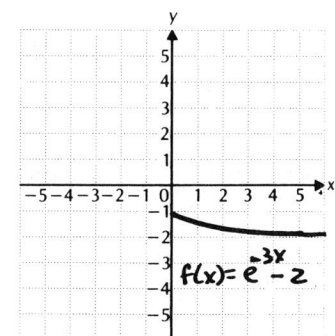

7. $\left(\sqrt{8}\right)^4 = 64$

8. $\log_a 0.00143 = -\dfrac{1}{3}$ 9. $5x + 2$ 10. x^3 11. 1.6340   12. -2, 4

13. $\log_a \dfrac{x^7 \sqrt[4]{z^3}}{\sqrt[3]{y}}$ 14. $\{x \mid x > 1\}$ 15. $\dfrac{1}{3}$ 16. $\dfrac{6}{13}$ 17. 6   18. $8\sqrt{10}$

19. 1.6371   20. -8.8116   21. Not defined   22. 4.8769   23. 0.544   24. 0.690

25. 1.623   26. (a) 12.2%; (b) 11.6%   27. (a) 182,211; (b) in 2021   28. About

3463 years   29. (a) 2.5 ft/sec; (b) 86,441

Chapter 5, Test Form F

1. b   2. d   3. c   4. b   5. a   6. c   7. d   8. b   9. c   10. c   11. a

12. d   13. b   14. c   15. d   16. b   17. b   18. a   19. c   20. b   21. d

22. a   23. c   24. b

1. (4,1)   2. Boat: 21 km/h; stream: 4 km/h   3. 21 black, 14 red   4. 5, 9, 15

5. $\left(\dfrac{1}{2}, -\dfrac{1}{4}\right)$ 6. $\left(\dfrac{2 + 5y}{4},\ y,\ \dfrac{3y - 6}{4}\right)$, (3,2,0), (-2,-2,-3), (8,6,3), etc. Answers

may vary.   7. (0,1,3)   8. Consistent, independent   9. Consistent, dependent

10. $f(x) = -x^2 + 2x + 3$ 11. 11   12. 27   13. (3,-2)   14. (1,-2,3)

15. Not possible   16. A or $\begin{bmatrix} -1 & 4 \\ 3 & 2 \end{bmatrix}$ 17. $\begin{bmatrix} 0 & 4 \\ 3 & 3 \end{bmatrix}$ 18. $\begin{bmatrix} 4 & 1 & -1 \\ 4 & 2 & 9 \\ 5 & 4 & 7 \end{bmatrix}$

19. $\begin{bmatrix} 0 & -1 \\ 1 & 5 \end{bmatrix}$ 20. $\begin{bmatrix} 5 & 8 \\ 23 & 18 \end{bmatrix}$ 21. Not possible   22. $\begin{bmatrix} -1 & 0 & 4 \end{bmatrix}$

23. $\begin{bmatrix} -1 & 5 & -8 \\ 8 & 1 & 6 \\ -5 & 2 & -4 \end{bmatrix}$ 24. $\begin{bmatrix} -1 & 5 \\ 2 & -3 \end{bmatrix}$ 25. $\begin{bmatrix} \dfrac{1}{5} & -\dfrac{2}{5} \\ \dfrac{1}{5} & \dfrac{3}{5} \end{bmatrix}$ 26. Does not exist

27. $\begin{bmatrix} -\dfrac{5}{3} & 1 & \dfrac{1}{3} \\ -\dfrac{13}{3} & 2 & \dfrac{2}{3} \\ -\dfrac{11}{3} & 2 & \dfrac{1}{3} \end{bmatrix}$ 28. $a_{12} = 2$; $M_{12} = \begin{vmatrix} 3 & 2 \\ -1 & 0 \end{vmatrix} = 2$; $A_{12} = -2$ 29. 10   30. 0

31. -8   32. 28   33. $(a - b)(c - b)(c - a)$ 34. $\begin{bmatrix} 3 & -2 \\ 1 & 3 \end{bmatrix}\begin{bmatrix} x \\ y \end{bmatrix} = \begin{bmatrix} -11 \\ 11 \end{bmatrix}$;

(-1,4)

35.

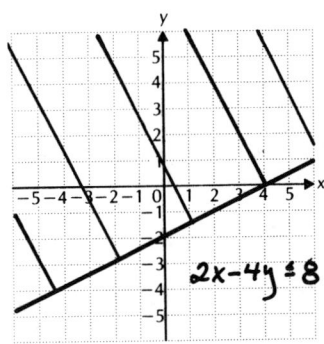

$2x - 4y \leq 8$

36. Minimum 20 at $(2,0)$; maximum $58\frac{1}{3}$

at $\left(4, \frac{11}{3}\right)$ 37. Bread: 300; cakes: 400;

maximum profit: $1800

38. $\dfrac{5}{3x - 1} - \dfrac{2}{x - 4}$ 39. $\left(-1, -\dfrac{1}{4}\right)$

Chapter 6, Test Form B

1. $(1, -4)$ 2. Airplane: 775 km/h; wind: 125 km/h   3. 24 L of 10%; 36 L of 35%

4. A = 70°, B = 62°, C = 48°   5. $\left(\dfrac{2}{3}, -\dfrac{1}{2}\right)$ 6. $(1, 2, 1)$

7. $(14 + 2z, 18 + 3z, z)$; $(14, 18, 0)$, $(16, 21, 1)$, $(10, 12, -2)$, etc. Answers may

vary.   8. Inconsistent, independent   9. Consistent, independent

10. $f(x) = -2x^2 - x + 1$ 11. -9 12. 28   13. $(-5, 4)$ 14. $(0, -3, 2)$

15. Not possible   16. $\begin{bmatrix} 3 & -2 \\ 4 & 1 \end{bmatrix}$ 17. $\begin{bmatrix} 4 & -2 \\ 4 & 2 \end{bmatrix}$ 18. $\begin{bmatrix} -1 & 7 & -3 \\ 6 & -1 & 0 \\ -1 & 9 & 2 \end{bmatrix}$

19. $\begin{bmatrix} 5 & -6 \\ 5 & -2 \end{bmatrix}$ 20. $\begin{bmatrix} -2 & 11 \\ 5 & 8 \end{bmatrix}$ 21. Not possible   22. $\begin{bmatrix} -2 & -1 & 3 \end{bmatrix}$

23. $\begin{bmatrix} 4 & 2 & -3 \\ 3 & -2 & 6 \\ 1 & 6 & 1 \end{bmatrix}$ 24. $\begin{bmatrix} -2 & 4 \\ -1 & 3 \end{bmatrix}$ 25. $\begin{bmatrix} \frac{2}{7} & \frac{1}{7} \\ -\frac{3}{7} & \frac{2}{7} \end{bmatrix}$ 26. $\begin{bmatrix} \frac{1}{25} & -\frac{4}{25} & \frac{12}{25} \\ \frac{6}{25} & \frac{1}{25} & -\frac{3}{25} \\ \frac{4}{25} & \frac{9}{25} & -\frac{2}{25} \end{bmatrix}$

27. Does not exist   28. $a_{23} = -1$; $M_{23} = \begin{vmatrix} 3 & -2 \\ 1 & -2 \end{vmatrix} = -4$; $A_{23} = 4$

29. 25   30. 12   31. 50   32. 1

Chapter 6, Test Form B (continued)

33. $(x - y)(z - y)(x - z)$ 34. $\begin{bmatrix} 3 & -1 \\ 1 & 2 \end{bmatrix}\begin{bmatrix} x \\ y \end{bmatrix} = \begin{bmatrix} 9 \\ -4 \end{bmatrix}$; $(2,-3)$

35.

36. Minimum -3 at $(-1,0)$; maximum 25 at

$\left(-1, \dfrac{7}{2}\right)$ 37. Type A: 7, Type B: 3;

maximum score: 85   38. $\dfrac{3}{4x - 3} - \dfrac{2}{x + 5}$

39. $m = \dfrac{1}{2}$; $b = -3$

Chapter 6, Test Form C

1. $(2,-1)$ 2. 5   3. Bill: 5; Ann: 10   4. $580 at 5%; $1000 at 8%;

$340 at 10%   5. $\left(-\dfrac{1}{3}, \dfrac{1}{4}\right)$ 6. $\left(\dfrac{19 - 7y}{13}, y, \dfrac{11 + 11y}{13}\right)$; $\left(\dfrac{19}{13}, 0, \dfrac{11}{13}\right)$, $\left(\dfrac{5}{13}, 2, \dfrac{33}{13}\right)$,

$(2,-1,0)$, etc. Answers may vary.   7. $(1,2,1)$ 8. Consistent, dependent

9. Consistent, independent   10. $f(x) = x^2 - 4x + 1$ 11. 8   12. -24

13. $(-1,-3)$ 14. $(-2,1,1)$ 15. Not possible   16. A or $\begin{bmatrix} -3 & 1 \\ 0 & 2 \end{bmatrix}$

17. $\begin{bmatrix} -2 & 1 \\ 0 & 3 \end{bmatrix}$ 18. $\begin{bmatrix} 7 & -2 & -1 \\ 8 & -4 & 10 \\ 2 & -1 & 5 \end{bmatrix}$ 19. $\begin{bmatrix} 0 & 2 \\ 2 & 6 \end{bmatrix}$ 20. $\begin{bmatrix} -4 & 4 \\ 13 & -14 \end{bmatrix}$

21. Not possible   22. $\begin{bmatrix} -5 & 2 & -7 \end{bmatrix}$ 23. $\begin{bmatrix} -4 & 2 & -5 \\ 4 & 1 & 5 \\ 4 & -8 & 7 \end{bmatrix}$

24. $\begin{bmatrix} -3 & -1 \\ -2 & -4 \end{bmatrix}$ 25. $\begin{bmatrix} -\dfrac{2}{5} & \dfrac{3}{10} \\ \dfrac{1}{5} & \dfrac{1}{10} \end{bmatrix}$

<u>26</u>. Does not exist   <u>27</u>. $\begin{bmatrix} \dfrac{9}{17} & -\dfrac{3}{17} & \dfrac{5}{17} \\ \dfrac{2}{17} & \dfrac{5}{17} & \dfrac{3}{17} \\ \dfrac{12}{17} & -\dfrac{4}{17} & \dfrac{1}{17} \end{bmatrix}$ <u>28</u>. $a_{31} = 2$;

$M_{31} = \begin{vmatrix} -3 & 2 \\ 5 & -2 \end{vmatrix} = -4$; $A_{31} = -4$ <u>29</u>. -14   <u>30</u>. 0   <u>31</u>. -30   <u>32</u>. 42

<u>33</u>. $(x - y)(z - y)(x - z)$ <u>34</u>. $\begin{bmatrix} 2 & -4 \\ 3 & 2 \end{bmatrix}\begin{bmatrix} x \\ y \end{bmatrix} = \begin{bmatrix} 6 \\ 1 \end{bmatrix}$; $(1,-1)$

<u>35</u>.

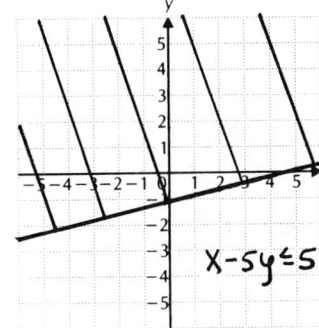

$X - 5y \leq 5$

<u>36</u>. Minimum -12 at $(0,-1)$; maximum 64 at $(2,4)$ <u>37</u>. 40 slacks, 160 skirts; maximum profit: \$2880

<u>38</u>. $\dfrac{5}{2x + 1} - \dfrac{3}{x - 4}$ <u>39</u>. $(1,-2,0,3)$

Chapter 6, Test Form D

<u>1</u>. $(-3,1)$ <u>2</u>. 1080 km   <u>3</u>. 11 dimes, 16 nickels   <u>4</u>. A: 350; B: 270; C: 420

<u>5</u>. $\left(-\dfrac{3}{4},\dfrac{1}{2}\right)$ <u>6</u>. $(1,-1,3)$ <u>7</u>. $\left(\dfrac{-4 - 3y}{2}, y, \dfrac{12 + 7y}{2}\right)$, $(-2,0,6)$, $\left(-\dfrac{7}{2}, 1, \dfrac{19}{2}\right)$,

$\left(-\dfrac{1}{2}, -1, \dfrac{5}{2}\right)$, etc. Answers may vary.   <u>8</u>. Consistent, independent   <u>9</u>. Consistent, dependent   <u>10</u>. $f(x) = 2x^2 - x - 3$ <u>11</u>. 12   <u>12</u>. 20   <u>13</u>. $(7,-2)$

<u>14</u>. $(5,0,-1)$ <u>15</u>. Not possible   <u>16</u>. A or $\begin{bmatrix} -2 & -3 \\ 1 & 4 \end{bmatrix}$ <u>17</u>. $\begin{bmatrix} -1 & -3 \\ 1 & 5 \end{bmatrix}$

18. $\begin{bmatrix} -2 & 4 & 4 \\ 3 & 1 & 0 \\ 10 & -1 & 7 \end{bmatrix}$ 19. $\begin{bmatrix} 0 & 0 \\ -3 & -2 \end{bmatrix}$ 20. $\begin{bmatrix} 14 & 2 \\ 2 & 26 \end{bmatrix}$ 21. Not possible

22. $\begin{bmatrix} 2 & -3 & -5 \end{bmatrix}$ 23. $\begin{bmatrix} 2 & -1 & 5 \\ -9 & 2 & 12 \\ 11 & 1 & -1 \end{bmatrix}$ 24. $\begin{bmatrix} -2 & -3 \\ 4 & 6 \end{bmatrix}$ 25. $\begin{bmatrix} -\dfrac{3}{14} & \dfrac{1}{14} \\ \dfrac{1}{7} & \dfrac{2}{7} \end{bmatrix}$

26. $\begin{bmatrix} -1 & 0 & -1 \\ 1 & \dfrac{1}{5} & \dfrac{2}{5} \\ -1 & \dfrac{1}{5} & -\dfrac{3}{5} \end{bmatrix}$ 27. Does not exist   28. $a_{23} = 5$;

$M_{23} = \begin{vmatrix} 4 & 0 \\ 2 & -3 \end{vmatrix} = -12$; $A_{23} = 12$ 29. -77   30. 0   31. -3   32. 14

33. $(a - b)(c - b)(c - a)$ 34. $\begin{bmatrix} 5 & -1 \\ 2 & 4 \end{bmatrix}\begin{bmatrix} x \\ y \end{bmatrix} = \begin{bmatrix} 11 \\ 22 \end{bmatrix}$; $(3,4)$

35.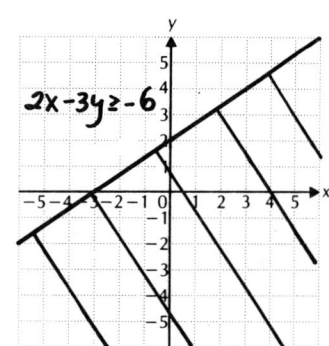

36. Minimum -36 at $(-2,-3)$; maximum 114 at $(8,-3)$ 37. 60 hamburgers; 40 hot dogs; maximum profit: $65

38. $\dfrac{7}{5x - 3} - \dfrac{2}{x + 4}$ 39. -2, 2

Chapter 6, Test Form E

1. $\left(\dfrac{1}{2}, -\dfrac{1}{2}\right)$ 2. $(5,-6)$ 3. $8500 at 6%, $6500 at 9%   4. 350   5. $(2y,y)$; $(0,0)$, $(2,1)$, $(-4,-2)$, etc. Answers may vary.   6. $(1,-2,1)$

Chapter 6, Test Form E (continued)

<u>7</u>. Consistent, independent   <u>8</u>. $f(x) = -2x^2 + 5x - 4$ <u>9</u>. -7   <u>10</u>. (3,-1,4)

<u>11</u>. 312   <u>12</u>. -108   <u>13</u>. 0   <u>14</u>. 270   <u>15</u>. $(x - y)(z - y)(z - x)$

<u>16</u>. $\begin{bmatrix} -3 & 4 \\ 2 & -5 \end{bmatrix} \begin{bmatrix} x \\ y \end{bmatrix} = \begin{bmatrix} 10 \\ -9 \end{bmatrix}$; (-2,1)   <u>17</u>. f   <u>18</u>. ℓ <u>19</u>. a   <u>20</u>. j   <u>21</u>. g

<u>22</u>. d   <u>23</u>. k   <u>24</u>.

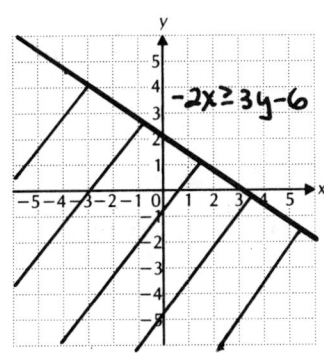

$-2x \geq 3y - 6$

<u>25</u>. Minimum 15 at (0,1); maximum $\dfrac{125}{3}$ at $\left(\dfrac{4}{3}, 1 \right)$

<u>26</u>. 3 dresses, 2 suits; maximum profit: $145

Chapter 6, Test Form F

<u>1</u>. a   <u>2</u>. b   <u>3</u>. b   <u>4</u>. d   <u>5</u>. d   <u>6</u>. d   <u>7</u>. b   <u>8</u>. c   <u>9</u>. d   <u>10</u>. c   <u>11</u>. a

<u>12</u>. d   <u>13</u>. b   <u>14</u>. c   <u>15</u>. d   <u>16</u>. a   <u>17</u>. d   <u>18</u>. d   <u>19</u>. a   <u>20</u>. d   <u>21</u>. a

<u>22</u>. b   <u>23</u>. b   <u>24</u>. b   <u>25</u>. d   <u>26</u>. c   <u>27</u>. b   <u>28</u>. d   <u>29</u>. a   <u>30</u>. a   <u>31</u>. b

<u>32</u>. a   <u>33</u>. c

1.

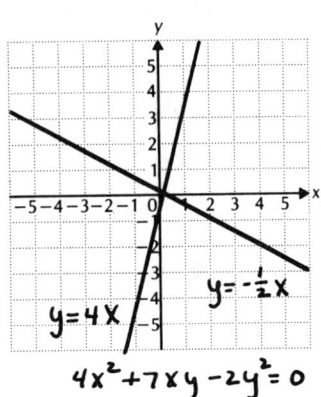

$y = -\frac{1}{2}x$

$y = 4x$

$4x^2 + 7xy - 2y^2 = 0$

2. C: (2,3); V: (3,3), (1,3), (2,6), (2,0);

F: $\left(2, 3 + 2\sqrt{2}\right)$, $\left(2, 3 - 2\sqrt{2}\right)$

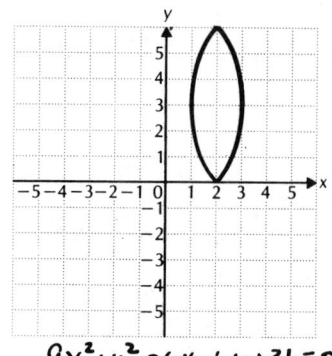

$9x^2 + y^2 - 36x - 6y + 36 = 0$

3. $\dfrac{x^2}{4} + \dfrac{y^2}{16} = 1$

4. C: (0,0); V: (0,4), (0,-4); F: $\left(0, 2\sqrt{5}\right)$, $\left(0, -2\sqrt{5}\right)$;

A: y = 2x, y = -2x

5.

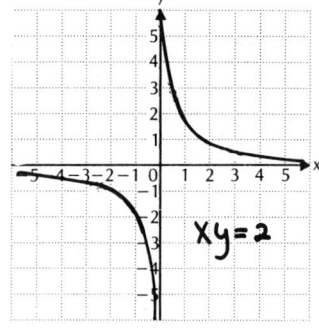

$xy = 2$

6. $y^2 = 24x$

7. V: (-1,3);

F: (-1,4); D: y = 2

8. (-8,6), (8,-6)   9. (1,1), (-1,-1)   10. 7 and 10   11. 9 ft by 12 ft

12. Numerator 11, denominator 10; or numerator 10, denominator 11

13. 2 ft by 12 ft   14. 4 and 18, or -4 and -18   15. 3 ft, 5 ft

16. Parabola   17. Hyperbola   18. Circle   19. Ellipse   20. Parabola

21. Hyperbola   22. $\dfrac{(x-3)^2}{4} + \dfrac{(y+1)^2}{9} = 1$ 23. $(x-2)^2 = 4(y+1)$

240

$\underline{24}$. $\left(\frac{1}{3},\frac{1}{6}\right)$, $\left(-\frac{1}{3},\frac{1}{6}\right)$, $\left(\frac{1}{3},-\frac{1}{6}\right)$, $\left(-\frac{1}{3},-\frac{1}{6}\right)$

Chapter 7, Test Form B

$\underline{1}$.

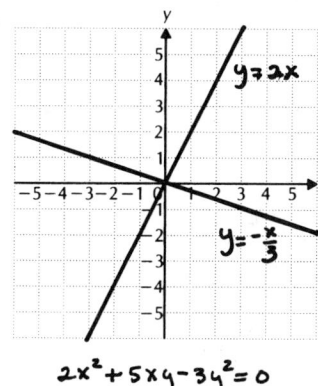

$y = 2x$

$y = -\frac{x}{3}$

$2x^2 + 5xy - 3y^2 = 0$

$\underline{2}$. C: $(1,-1)$; V: $(5,-1)$, $(-3,-1)$, $(1,-3)$, $(1,1)$;

F: $\left(1 + 2\sqrt{3},-1\right)$, $\left(1 - 2\sqrt{3},-1\right)$

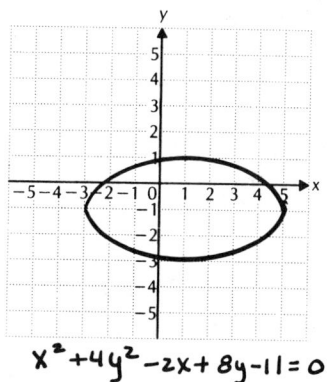

$x^2 + 4y^2 - 2x + 8y - 11 = 0$

$\underline{3}$. $\frac{x^2}{25} + y^2 = 1$ $\underline{4}$. C: $(0,0)$; V: $(3,0)$, $(-3,0)$; F: $(5,0)$, $(-5,0)$;

A: $y = \frac{4}{3}x$, $y = -\frac{4}{3}x$ $\underline{5}$.

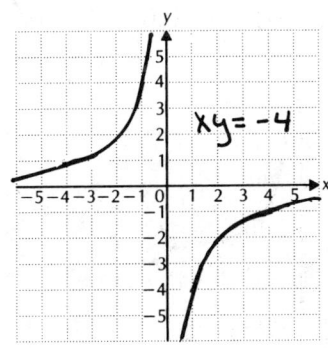

$xy = -4$

$\underline{6}$. $x^2 = 32y$

$\underline{7}$. V: $(4,-7)$; F: $(3,-7)$; D: $x = 5$ $\underline{8}$. $(0,4)$ $\underline{9}$. $(5,0)$, $(-4,3)$, $(-4,-3)$

$\underline{10}$. 10 and 15 $\underline{11}$. 5 cm by 7cm $\underline{12}$. Numerator 9, denominator 5;

or numerator 5, denominator 9 $\underline{13}$. 5 by 9 $\underline{14}$. 6 and 11, or -6 and -11

15. 7 ft, 8 ft   16. Circle   17. Ellipse   18. Parabola   19. Hyperbola

20. Parabola   21. Hyperbola   22. $\dfrac{x^2}{16} - \dfrac{y^2}{9} = 1$ 23. 2 and 5

24. $(x - 1)^2 + (y + 2)^2 = 1$

Chapter 7, Test Form C

1.

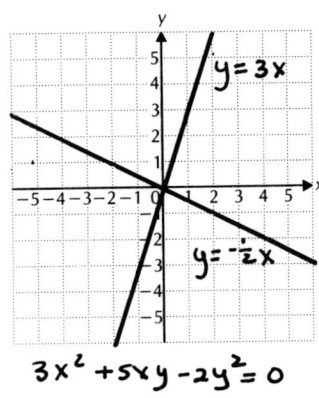

$3x^2 + 5xy - 2y^2 = 0$

2. C: (2,−3); V: (5,−3), (−1,−3), (2,−1), (2,−5);

F: $\left(2 + \sqrt{5}, -3\right)$, $\left(2 - \sqrt{5}, -3\right)$

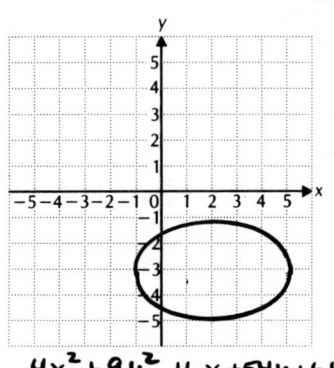

$4x^2 + 9y^2 - 16x + 54y + 61 = 0$

3. $\dfrac{x^2}{36} + \dfrac{y^2}{25} = 1$ 4. C: (0,0); V: (2,0), (−2,0); F: $\left(\sqrt{13}, 0\right)$, $\left(-\sqrt{13}, 0\right)$;

A: $y = \dfrac{3}{2}x$, $y = -\dfrac{3}{2}x$ 5.

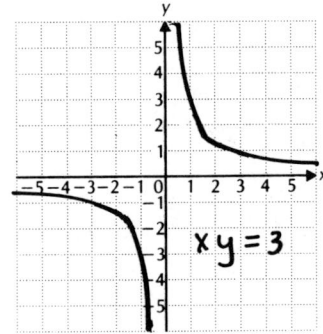

$xy = 3$

6. $y^2 = -12x$

Chapter 7, Test Form C (continued)

7. V: (1,-5); F: (1,-6); D: y = -4   8. (-1,2), (-4,1)   9. (2,2), (2,-2),

(-2,2), (-2,-2)   10. 3 and 12   11. 4 in. by 8 in.   12. Numerator 8,

denominator 13; or numerator 13, denominator 8   13. 5 yd by 12 yd

14. 5 and 15; or -5 and -15   15. 2 cm, 7 cm   16. Circle   17. Parabola

18. Hyperbola   19. Ellipse   20. Hyperbola   21. Parabola

22. $\dfrac{(x + 2)^2}{25} + \dfrac{(y - 3)^2}{16} = 1$ 23. 3 and 4   24. $(x + 3)^2 + (y - 2)^2 = 1$

Chapter 7, Test Form D

1.
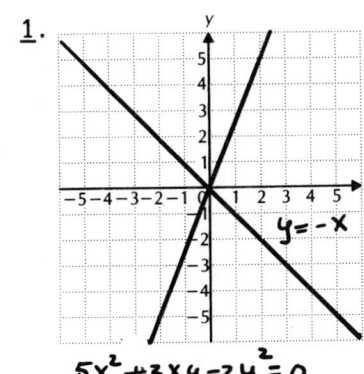

$5x^2 + 3xy - 2y^2 = 0$

2. C: (2,1); V: (6,1), (-2,1), (2,4), (2,-2);

F: $\left(2 + \sqrt{7}, 1\right)$, $\left(2 - \sqrt{7}, 1\right)$

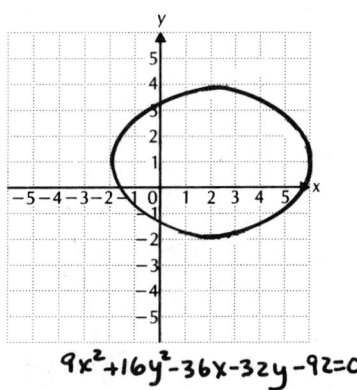

$9x^2 + 16y^2 - 36x - 32y - 92 = 0$

3. $\dfrac{x^2}{16} + \dfrac{y^2}{9} = 1$ 4. C: (0,0); V: (0,3), (0,-3); F: $\left(0, \sqrt{34}\right)$, $\left(0, -\sqrt{34}\right)$;

A: $y = \dfrac{3}{5}x$, $y = -\dfrac{3}{5}x$

5.

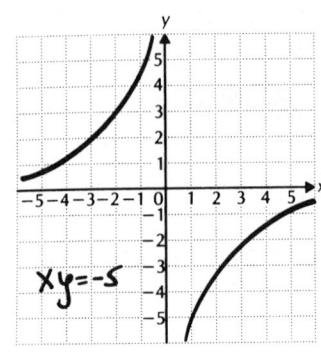

6. $x^2 = -24y$ 7. V: (2,3); F: (3,3); D: x = 1

8. (8,2), (5,1)   9. (3,3), (-3,-3), (9,1), (-9,-1)

10. 4 and 9   11. 40 yd by 115 yd   12. Numerator 7,

denominator 16; or numerator 16, denominator 7

13. 4 by 15   14. 6 and 16; or -6 and -16

15. 4 ft, 11 ft   16. Hyperbola   17. Parabola

18. Ellipse   19. Hyperbola   20. Parabola   21. Circle   22. $\frac{y^2}{4} - \frac{x^2}{9} = 1$

23. $(y - 2)^2 = 4(x + 3)$ 24. $\frac{5}{7}$

Chapter 7, Test Form E

1. ℓ 2. d   3. n   4. b   5. j   6. i   7. Parabola   8. Hyperbola

9. Circle   10. Ellipse   11.   12.

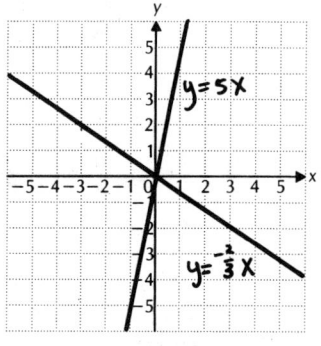

$10x^2 + 13xy - 3y^2 = 0$

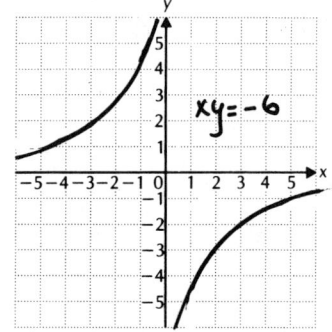

13.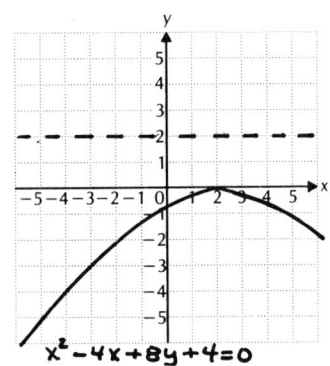

$x^2 - 4x + 8y + 4 = 0$

14.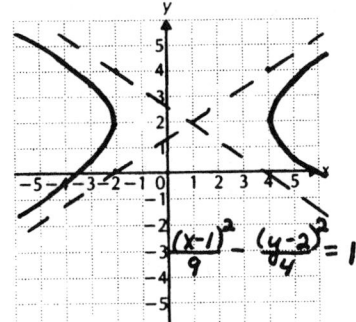

$\dfrac{(x-1)^2}{9} - \dfrac{(y-2)^2}{4} = 1$

15. $(4,0)$, $(0,2)$

16. $(2,2)$, $(2,-2)$, $(-2,2)$, $(-2,-2)$ 17. 6 and -12, or -6 and 12

18. $(y + 4)^2 = -5(x - 3)$

Chapter 7, Test Form F

1. c   2. a   3. b   4. b   5. d   6. b   7. b   8. d   9. a   10. c   11. a

12. b   13. b   14. a   15 c

Chapter 8, Test Form A

1. 7   2. 4   3. 22nd   4. $\dfrac{27}{4}, \dfrac{17}{2}, \dfrac{41}{4}$ 5. $-\dfrac{1}{1024}$ 6. $\dfrac{63}{8}$ 7. b   8. $\dfrac{216}{7}$

9. \$305, \$4800   10. $\dfrac{1}{81}$ ft   11. $\dfrac{8}{45}$

12. S_n: $1\cdot2 + 2\cdot4 + 3\cdot8 + \ldots + n\cdot2^n = (n - 1)2^{n+1} + 2$

S_1: $2 = (1 - 1)2^{1+1} + 2$

S_k: $1\cdot2 + 2\cdot4 + 3\cdot8 + \ldots + k\cdot2^k = (k - 1)2^{k+1} + 2$

S_{k+1}: $1\cdot2 + 2\cdot4 + 3\cdot8 + \ldots + k\cdot2^k + (k + 1)2^{k+1} = (k + 1 - 1)2^{k+1+1} + 2$
$$= k\cdot2^{k+2} + 2$$

1. Basis step: $(1 - 1)2^{1+1} + 2 = 2$, so S_1 is true.

2. Induction step: Assume S_k. Then add $(k + 1)\cdot2^{k+1}$ on both sides.

$1\cdot2 + 2\cdot4 + 3\cdot8 + \ldots + k\cdot2^k + (k + 1)2^{k+1} = (k - 1)2^{k+1} + 2 +$
$$(k + 1)2^{k+1}$$
$$= 2^{k+1}(k - 1 + k + 1) + 2$$
$$= 2^{k+1}(2k) + 2$$
$$= 2^{k+2}\cdot k + 2$$
$$= k\cdot2^{k+2} + 2$$

13. 3, 10, 101, 10,202   14. (a) 7^4, or 2401; (b) 840; (c) 210   15. 462

16. 4704   17. 2520   18. 2^7, or 128   19. $-540a^3b^3$

20. $x^4 + 4\sqrt{3}x^3 + 18x^2 + 12\sqrt{3}x + 9$ 21. $\frac{1}{9}$ 22. $\frac{2}{13}$ 23. $\frac{420}{4199}$ 24. 3, 6.25,

12.703704, 25.628906, 51.53632   25. 45

Chapter 8, Test Form B

1. 14   2. 7   3. 10th   4. $\frac{39}{4}, \frac{27}{2}, \frac{69}{4}$ 5. 9.6   6. $\frac{211}{81}$ 7. b, c

8. $\frac{3}{2}$ 9. 5184 ft   10. $1464.10   11. $\frac{8}{33}$

12. $S_n = \dfrac{1}{1 \cdot 2 \cdot 3} + \dfrac{1}{2 \cdot 3 \cdot 4} + \cdots + \dfrac{1}{n(n+1)(n+2)} = \dfrac{n(n+3)}{4(n+1)(n+2)}$

$S_1 = \dfrac{1}{1 \cdot 2 \cdot 3} = \dfrac{1(4)}{4(2)(3)}$

$S_k = \dfrac{1}{1 \cdot 2 \cdot 3} + \dfrac{1}{2 \cdot 3 \cdot 4} + \cdots + \dfrac{1}{k(k+1)(k+2)} = \dfrac{k(k+3)}{4(k+1)(k+2)}$

$S_{k+1} = \dfrac{1}{1 \cdot 2 \cdot 3} + \dfrac{1}{2 \cdot 3 \cdot 4} + \cdots + \dfrac{1}{k(k+1)(k+2)} + \dfrac{1}{(k+1)(k+2)(k+3)} = \dfrac{(k+1)(k+4)}{4(k+2)(k+3)}$

1. Basis step: $\dfrac{1}{6} = \dfrac{4}{24}$, so S_1 is true

2. Induction step: Assume S_k. Then add $\dfrac{1}{(k+1)(k+2)(k+3)}$ on both
   sides.

$\dfrac{1}{1 \cdot 2 \cdot 3} + \dfrac{1}{2 \cdot 3 \cdot 4} + \cdots + \dfrac{1}{k(k+1)(k+2)} + \dfrac{1}{(k+1)(k+2)(k+3)}$

$= \dfrac{k(k+3)}{4(k+1)(k+2)} + \dfrac{1}{(k+1)(k+2)(k+3)}$

$= \dfrac{k(k+3)(k+3) + 4}{4(k+1)(k+2)(k+3)}$

$= \dfrac{k^3 + 6k^2 + 9k + 4}{4(k+1)(k+2)(k+3)}$

$= \dfrac{(k+1)(k+1)(k+4)}{4(k+1)(k+2)(k+3)}$

$= \dfrac{(k+1)(k+4)}{4(k+2)(k+3)}$

13. -4, 17, -25, 59   14. (a) 8^3, or 512; (b) 336; (c) 6   15. 495   16. 1120

17. 5040   18. 2^9, or 512   19. $21,504a^5b^2$ 20. $25\sqrt{5} - 125a + 50\sqrt{5}a^2 - 50a^3$

$+ 5\sqrt{5}a^4 - a^5$ 21. $\frac{1}{9}$ 22. $\frac{1}{4}$ 23. $\frac{105}{286}$ 24. \$2600, \$1690, \$1098.50, \$714.03,

\$464.12   25. 9

Chapter 8, Test Form C

1. 4   2. 8   3. 13th   4. $\frac{27}{4}, \frac{19}{2}, \frac{49}{4}$ 5. $\frac{3}{32}$ 6. -47   7. a   8. $\frac{70}{9}$

9. 127.4   10. About 29,387   11. $\frac{77}{9}$

12. S_n: $\dfrac{1}{3} + \dfrac{1}{15} + \dfrac{1}{35} + \ldots + \dfrac{1}{4n^2 - 1} = \dfrac{n}{2n + 1}$

S_1: $\dfrac{1}{3} = \dfrac{1}{2(1) + 1}$

S_k: $\dfrac{1}{3} + \dfrac{1}{15} + \dfrac{1}{35} + \ldots + \dfrac{1}{4k^2 - 1} = \dfrac{k}{2k + 1}$

S_{k+1}: $\dfrac{1}{3} + \dfrac{1}{15} + \dfrac{1}{35} + \ldots + \dfrac{1}{4k^2 - 1} + \dfrac{1}{4(k + 1)^2 - 1} = \dfrac{k + 1}{2(k + 1) + 1}$

1. Basis step: $\dfrac{1}{3} = \dfrac{1}{3}$, so S_1 is true.

2. Induction step: Assume S_k. Then add $\dfrac{1}{4(k + 1)^2 - 1}$ on both sides.

$\dfrac{1}{3} + \dfrac{1}{15} + \dfrac{1}{35} + \ldots + \dfrac{1}{4k^2 - 1} + \dfrac{1}{4(k + 1)^2 - 1} = \dfrac{k}{2k + 1} + \dfrac{1}{4(k + 1)^2 - 1}$

$= \dfrac{k}{2k + 1} + \dfrac{1}{4(k^2 + 2k + 1) - 1}$

$= \dfrac{k}{2k + 1} + \dfrac{1}{4k^2 + 8k + 3}$

$= \dfrac{k}{2k + 1} + \dfrac{1}{(2k + 1)(2k + 3)}$

$= \dfrac{k(2k + 3) + 1}{(2k + 1)(2k + 3)}$

$= \dfrac{2k^2 + 3k + 1}{(2k + 1)(2k + 3)}$

$= \dfrac{(2k + 1)(k + 1)}{(2k + 1)(2k + 3)}$

$= \dfrac{k + 1}{2k + 3}$

$= \dfrac{k + 1}{2(k + 1) + 1}$

13. 4, 5, 7, 11   14. (a) 6^5, or 7776; (b) 720; (c) 6   15. 4845   16. 199,584

17. 1260   18. 2^5, or 32   19. $-1458ab^5$ 20. $343 + 294\sqrt{7a} + 735a^2 +$

$140\sqrt{7a^3} + 105a^4 + 6\sqrt{7a^5} + a^6$ 21. $\frac{1}{6}$ 22. $\frac{2}{13}$ 23. $\frac{175}{2431}$ 24. 2, 5, 8, 11

25. 10

Chapter 8, Test Form D

1. 12   2. 5   3. 27th   4. $\frac{27}{4}, \frac{21}{2}, \frac{57}{4}$ 5. $\frac{4}{729}$ 6. $\frac{312}{125}$ 7. a, c   8. $-\frac{32}{5}$

9. 3550   10. $16,383   11. $\frac{60}{11}$

12. S_n: $1^3 + 3^3 + 5^3 + \ldots + (2n - 1)^3 = n^2(2n^2 - 1)$

   S_1: $1^3 = 1^2(2(1)^2 - 1)$

   S_k: $1^3 + 3^3 + 5^3 + \ldots + (2k - 1)^3 = k^2(2k^2 - 1)$

   S_{k+1}: $1^3 + 3^3 + 5^3 + \ldots + (2k - 1)^3 + [2(k + 1) - 1]^3 =$
   $$(k + 1)^2[2(k + 1)^2 - 1]$$

   1. Basis step: $1^3 = 1^2(2(1)^2 - 1)$, so S_1 is true.

   2. Induction step: Assume S_k. Then add $[2(k + 1) - 1]^3$ on both sides.

   $1^3 + 3^3 + 5^3 + \ldots + (2k - 1)^3 + [2(k + 1) - 1]^3 =$

   $$= k^2(2k^2 - 1) + [2(k + 1) - 1]^3$$
   $$= 2k^4 - k^2 + 8k^3 + 12k^2 + 6k + 1$$
   $$= 2k^4 + 8k^3 + 11k^2 + 6k + 1$$
   $$= (k^2 + 2k + 1)(2k^2 + 4k + 1)$$
   $$= (k + 1)^2[2(k + 1)^2 - 1]$$

13. 4, 18, 88, 438   14. (a) 9^4, or 6561; (b) 3024; (c) 42   15. 792   16. 160

17. 840   18. 2^6, or 64   19. $720x^3y^2$ 20. $x^5 - 5\sqrt{6x^4} + 60x^3 - 60\sqrt{6x^2} +$

$180x - 36\sqrt{6}$ 21. $\frac{1}{12}$ 22. $\frac{1}{2}$ 23. $\frac{90}{221}$ 24. 6, 4, 2   25. 5

Chapter 8, Test Form E

<u>1</u>. 285   <u>2</u>. 57   <u>3</u>. 50th   <u>4</u>. 10q + 15p   <u>5</u>. 145.8   <u>6</u>. $\dfrac{133}{1944}$ <u>7</u>. $\dfrac{31}{4}$

<u>8</u>. $\dfrac{26}{3}$, $\dfrac{34}{3}$, 14, $\dfrac{50}{3}$, $\dfrac{58}{3}$ <u>9</u>. \$2912   <u>10</u>. 650 m   <u>11</u>. 386,428   <u>12</u>. $\dfrac{181}{45}$

<u>13</u>. S_n: $1 + 3 + 6 + \ldots + \dfrac{1}{2}n(n + 1) = \dfrac{n(n + 1)(n + 2)}{6}$

S_1: $1 = \dfrac{1(1 + 1)(1 + 2)}{6}$

S_k: $1 + 3 + 6 + \ldots + \dfrac{1}{2}k(k + 1) = \dfrac{k(k + 1)(k + 2)}{6}$

S_{k+1}: $1 + 3 + 6 + \ldots + \dfrac{1}{2}k(k + 1) + \dfrac{1}{2}(k + 1)(k + 2) =$

$$\dfrac{(k + 1)(k + 2)(k + 3)}{6}$$

1. Basis step: $\dfrac{1(1 + 1)(1 + 2)}{6} = 1$, so S_1 is true.

2. Induction step: Assume S_k. Then add $\dfrac{1}{2}(k + 1)(k + 2)$ on both sides.

$1 + 3 + 6 + \ldots + \dfrac{1}{2}k(k + 1) + \dfrac{1}{2}(k + 1)(k + 2) =$

$$= \dfrac{k(k + 1)(k + 2)}{6} + \dfrac{1}{2}(k + 1)(k + 2)$$
$$= \dfrac{k(k + 1)(k + 2)}{6} + \dfrac{3(k + 1)(k + 2)}{6}$$
$$= \dfrac{k(k + 1)(k + 2) + 3(k + 1)(k + 2)}{6}$$
$$= \dfrac{(k + 1)(k + 2)(k + 3)}{6}$$

<u>14</u>. 480   <u>15</u>. 7!, or 5040; 6!, or 720   <u>16</u>. 1260   <u>17</u>. (a) 5^4, or 625; (b) 120;

(c) 72   <u>18</u>. 56   <u>19</u>. $17,500x^3y^4$ <u>20</u>. $32x^5 + 80x^4y^2 + 80x^3y^4 + 40x^2y^6 +$

$10xy^8 + y^{10}$ <u>21</u>. $\dfrac{1}{18}$ <u>22</u>. $\dfrac{1}{221}$ <u>23</u>. $\dfrac{77}{207}$

Chapter 8, Test Form F

<u>1</u>. a   <u>2</u>. c   <u>3</u>. d   <u>4</u>. b   <u>5</u>. a   <u>6</u>. b   <u>7</u>. c   <u>8</u>. d   <u>9</u>. c   <u>10</u>. a   <u>11</u>. b

12. S_n: $2 + 6 + 18 + \ldots + 2(3^{n-1}) = 3^n - 1$

   S_1: $2 = 3^1 - 1$

   S_k: $2 + 6 + 18 + \ldots + 2(3^{k-1}) = 3^k - 1$

   S_{k+1}: $2 + 6 + 18 + \ldots + 2(3^{k-1}) + 2(3^k) = 3^{k+1} - 1$

   1. Basis step: $3^1 - 1 = 2$, so S_1 is true.

   2. Induction step: Assume S_k. Then add $2(3^k)$ on both sides.

$$2 + 6 + 18 + \ldots + 2(3^{k-1}) + 2(3^k) = 3^k - 1 + 2(3^k)$$
$$= 3(3^k) - 1$$
$$= 3^{k+1} - 1$$

13. c   14. d   15. c   16. b   17. d   18. a   19. c   20. b   21. d   22. b

23. d   24. c   25. d

Final Exam, Test Form A

1. 15, 0   2. $-\sqrt[5]{8}$, $\sqrt{40}$ 3. 15   4. 15, 0, −18   5. 1.333×10^{-5}

6. 3.2×10^{-5} 7. $\dfrac{x^2 + 6x + 12}{(x - 2)(x + 2)(x + 3)}$ 8. $\{x \mid x \leq -\frac{13}{5}\}$ 9. $\left\{-\frac{1}{2}, \frac{3}{7}\right\}$

10. $x^2 + \sqrt{3}x - 6 = 0$ 11. $y = \dfrac{2400xz}{w^2}$ 12. $\{x \mid x \leq 9\}$ 13. $y = -4x - 5$

14. $\sqrt{13}$, or 3.6056   15. $(f + g)(x) = x^2 + 2x - 1$, $(f - g)(x) = -x^2 + 2x - 7$,

$fg(x) = 2x^3 - 4x^2 + 6x - 12$, $(f/g)(x) = \dfrac{2x - 4}{x^2 + 3}$, $f \circ g(x) = 2x^2 + 2$,

$g \circ f(x) = 4x^2 - 16x + 19$ 16. All reals, all reals, all reals, all reals,

all reals, all reals, all reals, all reals   17. c   18. a and b

19. $\{x \mid -1 \leq x \leq 6\}$ 20. $\{x \mid -\frac{7}{2} < x < \frac{1}{3}\}$ 21. b = 15 cm, h = 15 cm

22. $(x + 3)(x - 1 + \sqrt{2})(x - 1 - \sqrt{2})$; -3, $1 - \sqrt{2}$, $1 + \sqrt{2}$

23. $(x - 2)(x + 3)(x + 1)^2(x - 4)^3$

24.

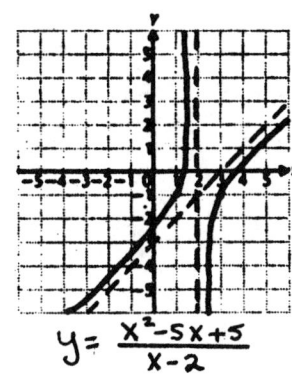

$$y = \frac{x^2 - 5x + 5}{x - 2}$$

25. $x = 2y - 5$ 26. 1.8502   27. 64   28. About 9 years   29. 4.1973   30. 0.8544   31. $(4,3)$

32. $(2,1,1)$ 33. $f(x) = -2x^2 + 3x - 4$ 34. 60

35. $\begin{bmatrix} 7 & -1 \\ 4 & 12 \end{bmatrix}$ 36. $\begin{bmatrix} \dfrac{2}{5} & -\dfrac{1}{5} \\ \dfrac{1}{15} & \dfrac{2}{15} \end{bmatrix}$ 37. $\dfrac{2}{x+1} + \dfrac{1}{x+2}$

38. $\dfrac{x^2}{4} + \dfrac{y^2}{9} = 1$

39.

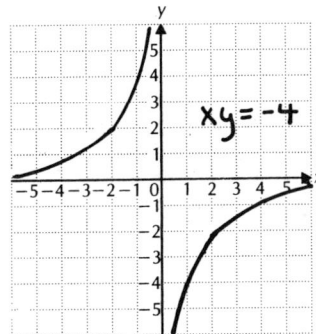

$xy = -4$

40. 3 and 11   41. 4 m × 15 m   42. Circle

43. 12,075   44. 150   45. (a) 9^5, or 59,049;

(b) 15,120; (c) 6561   46. 3, 7, 23, 87, 343

47. $x^5 + 5\sqrt{2}x^4 + 20x^3 + 20\sqrt{2}x^2 + 20x + 4\sqrt{2}$

48. $\dfrac{20}{221}$

Final Exam, Test Form B

1. −9.4   2. 70   3. 68   4. −28   5. $(2x - 5)(3x + 4)$

6. $(a + 7)(a^2 - 7a + 49)$ 7. $(x + 2y)(6 - x)$ 8. $\dfrac{2}{13} - \dfrac{3}{13}i$ 9. $\left\{-\dfrac{9}{4}, \dfrac{7}{3}\right\}$

10. 15   11. 6 in. × 18 in.   12. Two real   13. No   14. {3, 4, 8}

15. {−2, 1, 7}   16. $(x - 5)^2 + (y + 1)^2 = 9$

17.

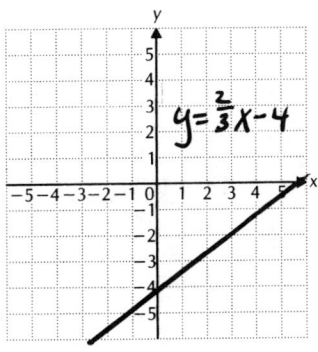

$$y = \frac{2}{3}x - 4$$

18. $(-\infty, -5]$

19.

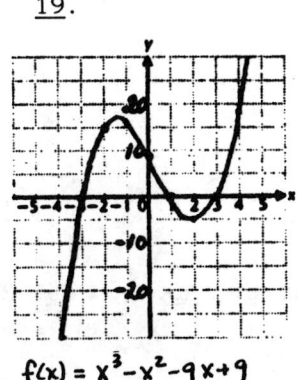

$$f(x) = x^3 - x^2 - 9x + 9$$

20. $\left(\frac{7}{2}, 0\right)$, $(-3,0)$ 21. 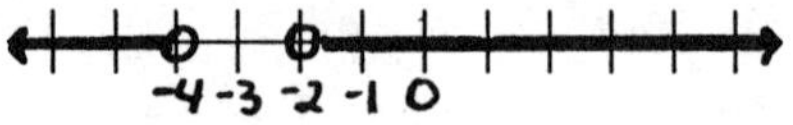

22. Yes

23. $\{1,2,3,5,7,9,13,15\}$ 24. $\{-7,4\}$ 25. a) $P(t) = 200{,}000e^{0.034t}$;

b) 280,989; c) 2010 26. 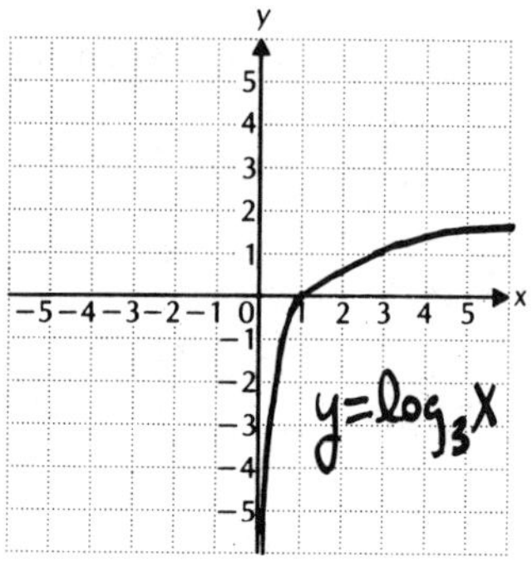

27. $\log_a \frac{x^4}{\sqrt{yz^2}}$

28. -8.2171

29. Not defined

30. 3.98×10^{-6}

31. $\frac{5}{3}$ km/h 32. $\left(\frac{8}{7}, \frac{11}{7}\right)$

33. 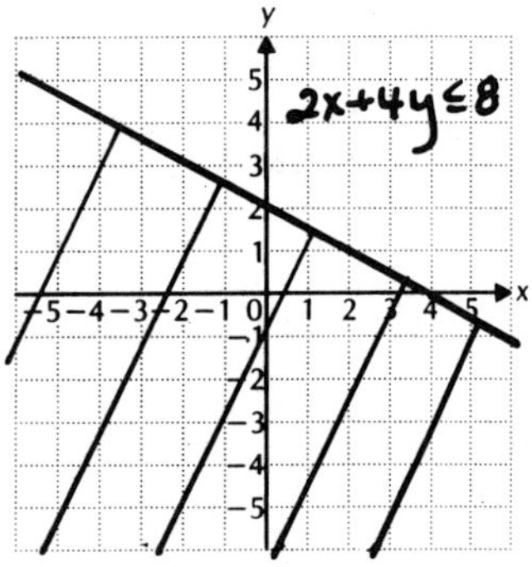

34. $a_{21} = 1$; $M_{21} = \begin{vmatrix} 0 & 4 \\ 3 & 7 \end{vmatrix} = -12$; $A_{21} = 12$

35. -21 36. $\begin{bmatrix} 33 & 5 \\ 9 & 6 \end{bmatrix}$ 37. Inconsistent, independent

38. 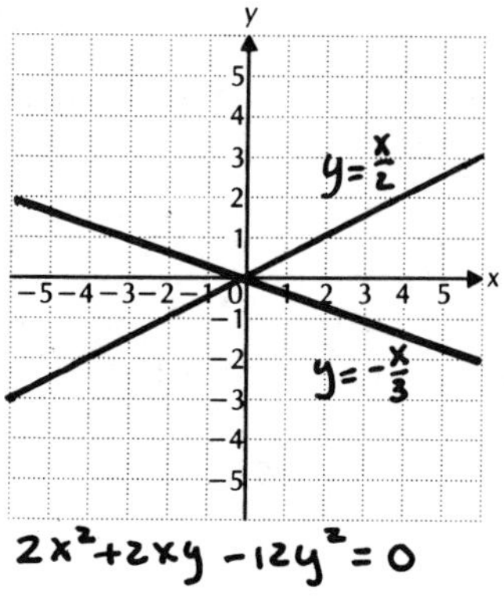

39. C: $(0,0)$; V: $(3,0)$, $(-3,0)$; F: $(5,0)$, $(-5,0)$; A: $y = \frac{4}{3}x$, $y = -\frac{4}{3}x$

40. $(1,2)$, $(6,3)$ 41. 11 and 12, -11 and -12 42. Hyperbola 43. 13th

44. a, c 45. $\frac{1}{6}$ 46. 2^8, or 256 47. $84{,}375x^2y^4$ 48. $\frac{2}{13}$

Final Exam, Test Form C

1. 0.00006129   2. 41,000   3. 2150   4. 0.0129   5. $\dfrac{(x - 5)(x + 3)}{(x + 1)(x + 2)}$

6. $\dfrac{49 + 14\sqrt{x} + x}{49 - x}$ 7. 11   8. $-i$ 9. $3 + 11i$ 10. $-\dfrac{5}{17} + \dfrac{14}{17}i$ 11. 50

12. 6 mph   13. b   14. d   15. a, b, c, e, f   16. 18   17. $2a^2 + 3a - 2$

18. $2h - 5 + 4a$

19.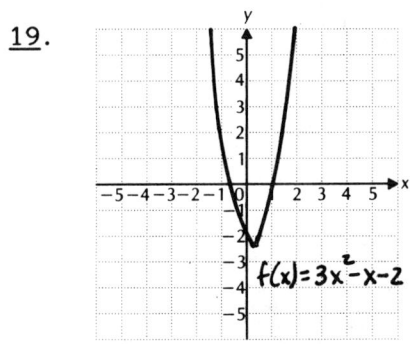

20. $\{x \mid x < 4 \text{ or } x > 12\}$ 21. $\{x \mid x < -7 \text{ or } x > 3\}$

22. $\{x \mid x < -3 \text{ or } x > -2\}$ 23. 93

24. $(x + 3)(x - 1 + \sqrt{2})(x - 1 - \sqrt{2})$; -3, $1 - \sqrt{2}$, $1 + \sqrt{2}$ 25. a and b   26. 3x   27. $\dfrac{5}{4}$ 28. 3

29. 3590 yrs   30. a) 80; b) 70; c) 45 months

31. 19   32. $(1 + z, -3z, z)$: $(1,0,0)$; $(3,-6,2)$; $(0,3,-1)$, etc. Answers may vary.

33. $\dfrac{49}{4}$ 34. $\begin{bmatrix} 5 & -1 \\ 1 & 3 \end{bmatrix} \begin{bmatrix} x \\ y \end{bmatrix} = \begin{bmatrix} 11 \\ -1 \end{bmatrix}$; $(2,-1)$ 35. Minimum -36 at $(-3,7)$;

maximum 0 at $(0,0)$ 36. $\begin{bmatrix} 4 & -5 \\ 2 & 9 \end{bmatrix}$ 37. $\begin{bmatrix} -\dfrac{3}{2} & 2 \\ -\dfrac{1}{2} & 1 \end{bmatrix}$ 38.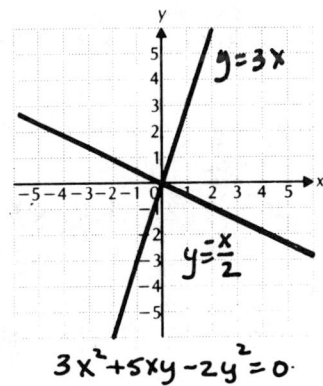

39. V: $(-\dfrac{1}{2}, 5)$; F: $(2, 5)$; D: $x = -3$ 40. $(3,-2)$,

$(-3,2)$, $\left(2\sqrt{3}, -\sqrt{3}\right)$, $\left(-2\sqrt{3}, \sqrt{3}\right)$ 41. 5 cm, 9 cm

42. Ellipse   43. 90   44. $\dfrac{29}{4}$, $\dfrac{19}{2}$, $\dfrac{47}{4}$ 45. $-\dfrac{20}{3}$

46. $\dfrac{269}{33}$ 47. 1120   48. $\dfrac{75}{364}$

Final Exam, Test Form D

<u>1</u>. 3.4×10^{-6} <u>2</u>. 8.9417×10^{2} <u>3</u>. $-12x^{3}y^{-2}$ <u>4</u>. -4 <u>5</u>. 1

<u>6</u>. $25a^{6} - 20a^{3}b^{2} + 4b^{4}$ <u>7</u>. 800 g/cm   <u>8</u>. $\frac{3}{2}$, 1, -5 <u>9</u>. $\{y \mid y > -2\}$

<u>10</u>. $\frac{11}{3}$ <u>11</u>. 4 <u>12</u>. $V_{2} = \dfrac{P_{1} V_{1} T_{2}}{P_{2} T_{1}}$ <u>13</u>.

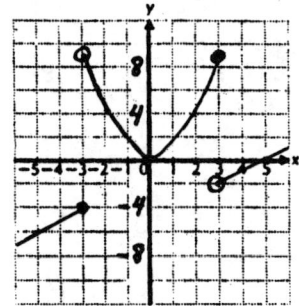

<u>14</u>. $P(x) = -0.5x^{2} + 12x - 360$

<u>15</u>. $m = -\frac{1}{2}$, y-intercept is $\frac{15}{4}$

<u>16</u>. $y = -\frac{1}{4}x + \frac{11}{4}$

<u>17</u>. Center: $(-3,5)$; radius: 4

<u>18</u>. a, c   <u>19</u>. a) $f(x) = 1\left(x - \frac{5}{2}\right)^{2} + \frac{1}{4}$; b) $\left(\frac{5}{2}, \frac{1}{4}\right)$, c) maximum: $\frac{1}{4}$

<u>20</u>. a) $f(x) = 2(x - 1)^{2} - 5$; b) $(1,-5)$; c) minimum: -5 <u>21</u>. $\{2, 12\}$

<u>22</u>. The quotient is $2x^{3} - 4x^{2} - 4x - 5$.   The remainder is 2.

<u>23</u>. $x^{3} - 3x^{2} + 4x - 12$ <u>24</u>. Positive: 3 or 1; negative: 2 or 0

<u>25</u>. $\{(-3,2), (-2,4), (5,1), (-7,-2), (-1.5,2.1)\}$ <u>26</u>. -5

<u>27</u>. $2 \log a + \log b - 5 \log c$ <u>28</u>. 39 decibels   <u>29</u>. 0.699   <u>30</u>. 0.954

<u>31</u>. $30°$ <u>32</u>. $(1,0,-3)$ <u>33</u>. A, or $\begin{bmatrix} 5 & -3 \\ 2 & 0 \end{bmatrix}$ <u>34</u>. A, or $\begin{bmatrix} 5 & -3 \\ 2 & 0 \end{bmatrix}$

<u>35</u>. $\begin{bmatrix} -3 & -2 & 4 \end{bmatrix}$ <u>36</u>. $\begin{bmatrix} 13 \end{bmatrix}$ <u>37</u>. 20 pies and 20 cakes   <u>38</u>. C: $(2,-3)$;
V: $(2,-1)$, $(2,-5)$, $(5,-3)$, $(-1,-3)$; F: $\left(2 - \sqrt{5}, -3\right)$, $\left(2 + \sqrt{5}, -3\right)$

<u>39</u>. $y^{2} = 16x$ <u>40</u>. 8 ft x 12 ft   <u>41</u>. $(2,1)$, $(2,-1)$, $(-2,1)$, $(-2,-1)$

<u>42</u>. Parabola   <u>43</u>. 209   <u>44</u>. 19   <u>45</u>. 6144   <u>46</u>. \$1023   <u>47</u>. 3024

<u>48</u>. $x^{4} - 4\sqrt{3}x^{3} + 18x^{2} - 12\sqrt{3}x + 9$

Final Exam, Test Form E

1. $\sqrt[60]{p^{20}q^{45}r^{24}}$ 2. $\dfrac{2x + 7\sqrt{xy} + 3y}{x - 9y}$ 3. $\dfrac{x^2 - 3x + 9}{x}$ 4. $5x^2 - 17x - 12 = 0$

5. 7   6. $y = \dfrac{3.75\ xz^2}{w}$ 7. $\dfrac{23}{2}$ 8. No   9. $\{x\,|\,x \geq 2,\ x \neq 3\}$

10. a) $f(x) = 2\left(x - \dfrac{3}{2}\right)^2 - \dfrac{7}{2}$; b) $\left(\dfrac{3}{2}, -\dfrac{7}{2}\right)$; c) minimum at $-\dfrac{7}{2}$

11. $\{x\,|\,x < \dfrac{3}{2}$ or $x > 6\}$ 12. $x\,|\,x < 2$ or $x > 3\}$ 13. a) 1.146; b) 0.954

14. -3.2651 15. 4   16. $(2,1,-3)$ 17. Maximum 125 at $(0,5)$

18. $f(x) = -x^2 + 3x - 4$ 19. -3 20. $\begin{bmatrix} \dfrac{1}{4} & -\dfrac{1}{2} & \dfrac{1}{2} \\[2mm] \dfrac{1}{8} & \dfrac{1}{4} & -\dfrac{1}{4} \\[2mm] \dfrac{1}{4} & \dfrac{1}{2} & \dfrac{1}{2} \end{bmatrix}$ 21. $\begin{bmatrix} -9 & 22 \\ 14 & -4 \end{bmatrix}$

22. $(0,1)$, $(0,-1)$ 23. $x^2 + y^2 = 25$ 24. $(0,5)$, $(0,-5)$ 25. one

26. $2i$, $-\sqrt{5}$, $3 + 2i$ 27. -1, 2   28. 6.1   29. $\dfrac{128}{5}$

30. S_n: $1 + 6 + 11 + \ldots + (5n - 4) = \dfrac{n(5n - 3)}{2}$

S_1: $1 = \dfrac{1(5\cdot 1 - 3)}{2}$

S_k: $1 + 6 + 11 + \ldots + (5k - 4) = \dfrac{k(5k - 3)}{2}$

S_{k+1}: $1 + 6 + 11 + \ldots + (5k - 4) + [5(k + 1) - 4] =$

$$\dfrac{(k + 1)[5(k + 1) - 3]}{2}$$

$1 + 6 + 11 + \ldots + (5k - 4) + (5k + 1) = \dfrac{(k + 1)(5k + 2)}{2}$

1. Basis step: $\dfrac{1(5 - 3)}{2} = 1$, so S_1 is true.

2. Induction step. Assume S_k. Then add $(5k + 1)$ on both sides.

$1 + 6 + 11 + \ldots + (5k - 4) + (5k + 1) = \dfrac{k(5k - 3)}{2} + (5k + 1)$

$$= \dfrac{k(5k - 3)}{2} + \dfrac{2(5k + 1)}{2}$$

$$= \dfrac{k(5k - 3) + 2(5k + 1)}{2}$$

$$= \dfrac{5k^2 - 3k + 10k + 2}{2}$$

$$= \dfrac{5k^2 + 7k + 2}{2}$$

$$= \dfrac{(k + 1)(5k + 2)}{2}$$

31. 120   32. $16x^4 - 32\sqrt{3}x^3y + 72x^2y^2 - 24\sqrt{3}xy^3 + 9y^4$ 33. $\dfrac{96}{323}$

34.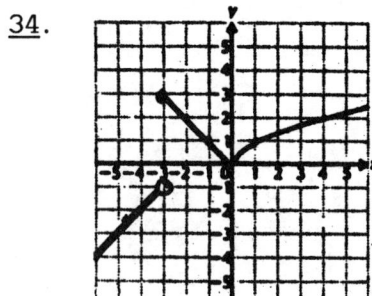

35.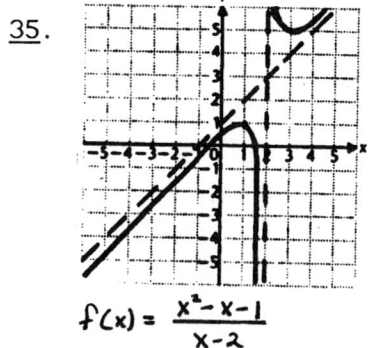

$f(x) = \dfrac{x^2 - x - 1}{x - 2}$

Final Exam, Test Form E (continued)

36.

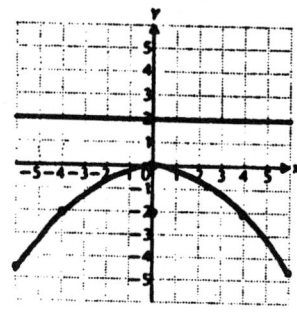

$$x^2 + 8y = 0$$

37. 11 yr 38. 78 39. 10 ft by 24 ft 40. -84 41. $\dfrac{4}{3x + 1} + \dfrac{1}{x + 1}$

Final Exam, Test Form F

1. b 2. a 3. a 4. b 5. d 6. b 7. c 8. c 9. b 10. d 11. a

12. c 13. a 14. a 15. d 16. c 17. c 18. a 19. d 20. b 21. c

22. a 23. a 24. d 25. a 26. d 27. a 28. d 29. c 30. b

31. S_n: $1 + 3 + 3^2 + \ldots + 3^{n-1} = \dfrac{3^n - 1}{2}$

S_1: $1 = \dfrac{3^1 - 1}{2}$

S_k: $1 + 3 + 3^2 + \ldots + 3^{k-1} = \dfrac{3^k - 1}{2}$

S_{k+1}: $1 + 3 + 3^2 + \ldots + 3^{k-1} + 3^k = \dfrac{3^{k+1} - 1}{2}$

1. Basis step: $\dfrac{3^1 - 1}{2} = 1$, so S_1 is true.

2. Induction step: Assume S_k. Then add 3^k on both sides.

$$1 + 3 + 3^2 + \ldots + 3^{k-1} + 3^k = \dfrac{3^k - 1}{2} + 3^k$$

$$= \dfrac{3^k - 1}{2} + \dfrac{2 \cdot 3^k}{2}$$

$$= \dfrac{3 \cdot 3^k - 1}{2}$$

$$= \dfrac{3^{k+1} - 1}{2}$$

32. a 33. a 34. d

ALGEBRA AND TRIGONOMETRY VIDEOTAPE SERIES

TAPE 1   11.7   The Binomial Theorem
39.39

TAPE 2   2.4   The Complex Numbers
39.30

TAPE 3   1.9   Handling Dimension Symbols
18.46

TAPE 4   3.1   Graphs of Equations
35.00

TAPE 5   3.3   Functions
35.00

TAPE 6   3.6   Symmetry
35.00   3.8   Transformations

TAPE 7   3.5   Some Special Classes of Functions
32.00

TAPE 8   3.4   Lines and Linear Functions
45.01   3.2, 3.4   Parallel and Perpendicular Lines; The Distance Formula

TAPE 9   4.1   Quadratic Functions
          4.1   Mathematical Models

TAPE 10   4.2   Sets, Sentences, and Inequalities
35.02   4.3   Equations and Inequalities with Absolute Value

TAPE 11   4.4   Quadratic and Rational Inequalities
21.49

TAPE 12   4.5, 4.6   Polynomials and Polynomial Functions
40.44   4.6   The Remainder and Factor Theorems

TAPE 13   4.7   Theorems about Roots
46.26   4.7   Rational Roots

TAPE 35   6.1   Systems of Equations in Two Variables
21.11

TAPE 36   6.2   Systems of Equations in Three or More Variables
32.36   6.3   Special Cases

TAPE 37   6.4   Matrices and Systems of Equations
29.55

TAPE 38   6.4   The Algebra of Matrices
30.09